Concerning C[illegible]
Adaptations & Interdependence

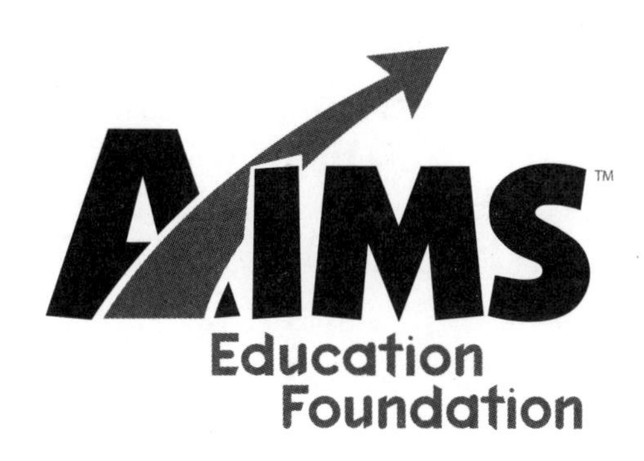

Developed and Published
by
AIMS Education Foundation

Research and Development
Jodi Bertolucci
Betty Cordel
Denise Del Grosso Gilliand
Sheldon Erickson
Suzy Gazlay
Judith Hillen
Burleigh Lockwood
David Mitchell
Myrna Mitchell
Michelle Pauls
Jim Wilson
Dave Youngs

Illustrators
Jeremy Balzer
Brock Heasley
Ben Hernandez
Renée Mason
Dawn McAndrews
Margo Pocock
David Schlotterback
Brenda Wood

Desktop Publisher
Tracey Lieder

Original Critters Writing Team
Maureen Murphy Allen
Debby Deal
Gale Philips Kahn
Suzanne Scheidt
Vincent Sipkovich

Concerning Critters:
Adaptations & Interdependence

Developed and Published
by
AIMS Education Foundation

This book contains materials developed by the AIMS Education Foundation. **AIMS** (**A**ctivities **I**ntegrating **M**athematics and **S**cience) began in 1981 with a grant from the National Science Foundation. The non-profit AIMS Education Foundation publishes hands-on instructional materials that build conceptual understanding. The foundation also sponsors a national program of professional development through which educators may gain expertise in teaching math and science.

AIMS Education Foundation
1595 S. Chuestnut Ave., Fresno, CA 93702-4706 • 888.733.2467 • aimsedu.org

ISBN 978-1-60519-077-8

Printed in the United States of America

Concerning Critters: Adaptations & Interdependence

Table of Contents

I Hear and I Forget,

I See and I Remember,

I Do and I Understand.

-Chinese Proverb

Assembling Rubber Band Books

Rubber band books offer valuable content information in a kid-friendly way. Each student can be given his or her own book to keep and refer to at a later date. These books also provide a great home link, as students can take them home and share the information they are learning with their parents. To assemble a book, follow these simple instructions:

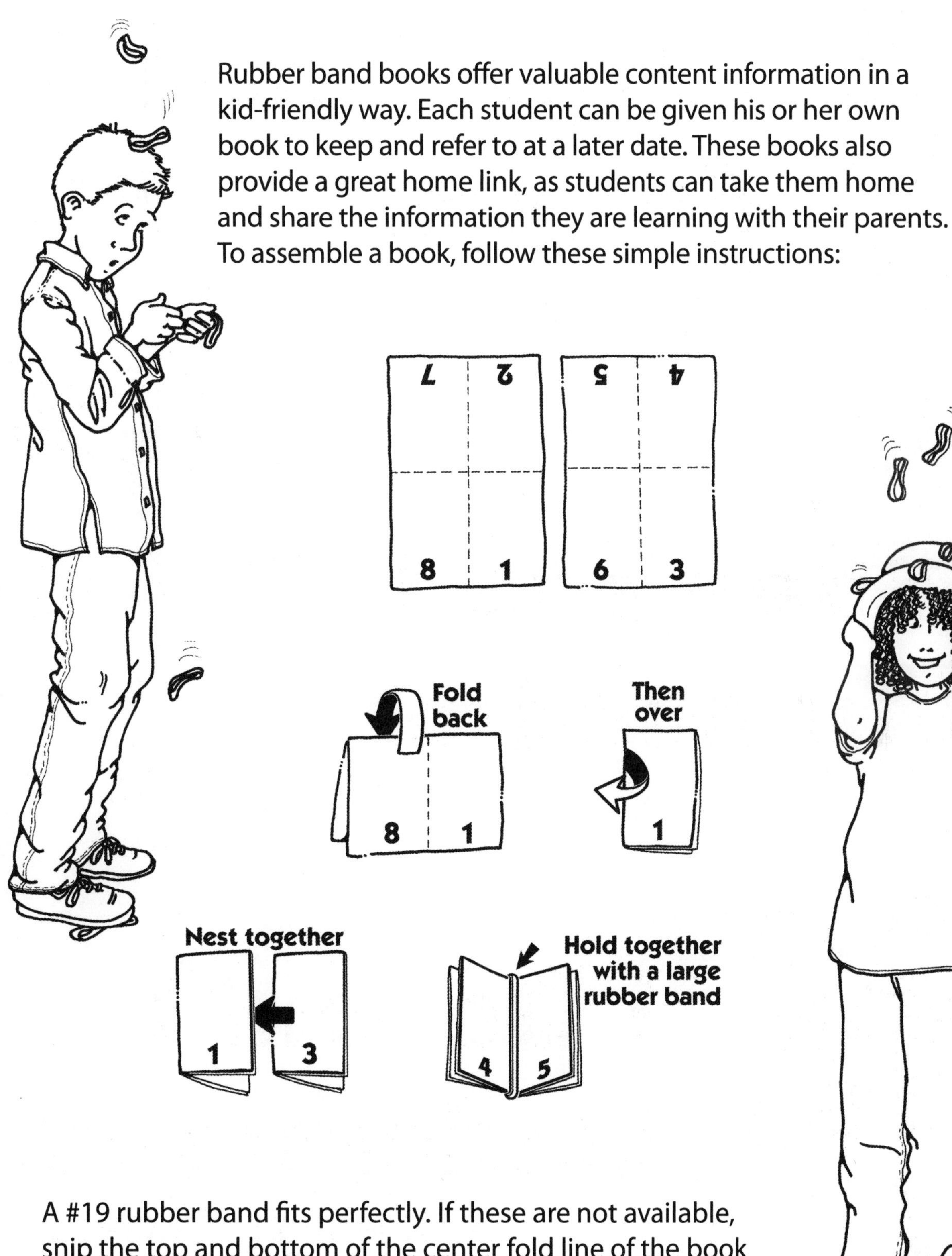

A #19 rubber band fits perfectly. If these are not available, snip the top and bottom of the center fold line of the book so that the other rubber bands can fit.

Adapting for Life

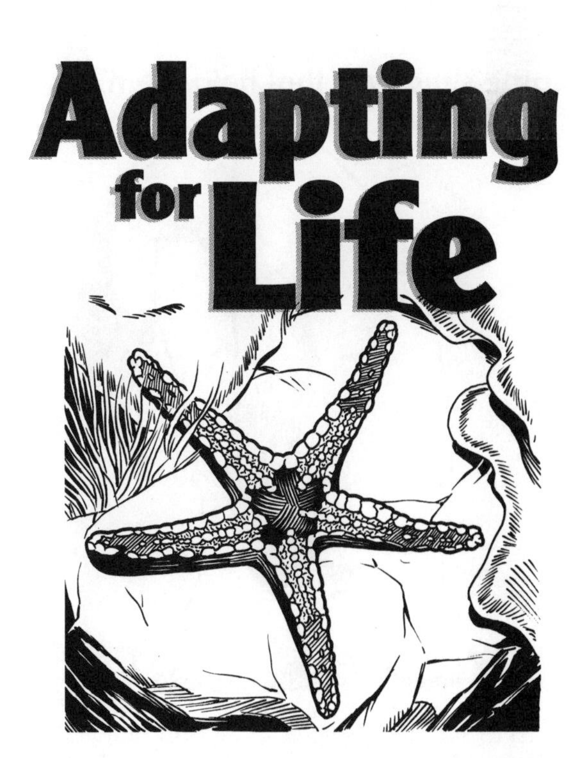

Animals have special adaptations that help them meet their needs.

1

Some have structures that help them gather food.

A catfish has barbels. These feel around on the bottom of the lake to help it find food.

2

A mother kangaroo has a special pouch for its baby. The baby feeds inside the pouch. It is also protected in the pouch.

A mother opossum also has a pouch that helps her care for her babies. Plus, she has a special type of tail that she uses to keep the babies from falling off her back.

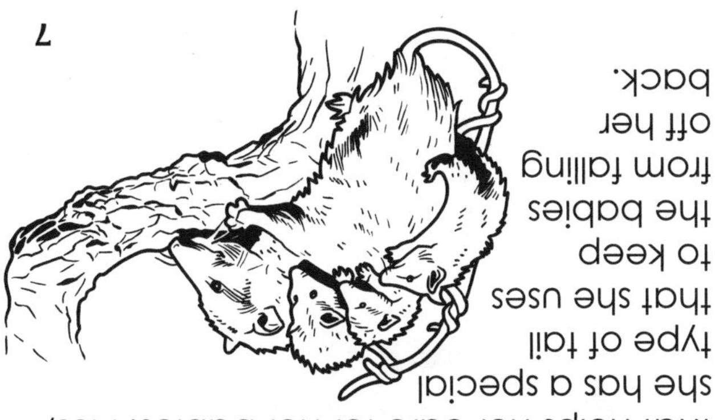

7

Some structures help them defend themselves.

A tarantula has tiny hairs on its abdomen. The hairs can stick the predator. They can cause itchy rashes.

A skunk has scent glands that spray a stinky liquid on its predators.

A deer will use its hoofs and antlers to defend itself.

8

A giraffe has a long neck that reaches high into the trees. Other grazing animals can't reach that high.

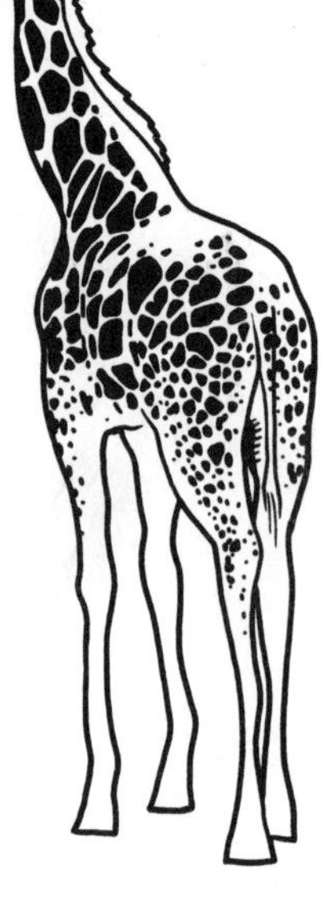

A pelican has a special beak. The beak has a built in pouch. The pelican uses the pouch as a dipping net to hold the fish it catches.

3

4

Some structures help them find shelter.

The beaver's front teeth help it cut down trees to build a dam.

5

A badger's front claws are about 3 centimeters (1 inch) long. They use their claws to dig their burrows.

A tent-making bat uses its teeth to build a tent out of a banana leaf. It lives in the rain forest. The tent protects it from the rain.

Some structures that help them raise their babies.

The father penguin has a flap of fat that covers the egg he carries on his feet. This flap of fat is like a blanket. It keeps the egg and the baby warm.

6

Table Manners

Topic
Adaptations

Key Question
How are insects adapted to eating certain types of food?

Learning Goals
Students will:
- simulate food gathering with four different types of insect mouths, and
- determine which mouth is most effective for different food sources.

Guiding Documents
Project 2061 Benchmarks
- *Different plants and animals have external features that help them thrive in different kinds of places.*
- *A model of something is different from the real thing but can be used to learn something about the real thing.*

NRC Standard
- *Each plant or animal has different structures that serve different functions in growth, survival, and reproduction. For example, humans have distinct body structures for walking, holding, seeing, and talking.*

Math
Measurement
  volume
Graphing

Science
Life science
  adaptations
  insect mouthparts

Integrated Processes
Observing
Predicting
Collecting and recording data
Classifying

Materials
Mouthparts (per group of four):
1 flexible drinking straw
1 regular straw with one end cut diagonally to form a point
1 clothespin
1 clothespin attached to a small piece of sponge about 1" x 1"

Food sources (per group of four):
paper torn and crumpled into small pieces
1 narrow-necked bottle containing water
1 cup containing water and covered with a paper towel or plastic wrap taped securely over its opening
1 tuna can or small bowl containing water

Collection apparatus (per group of four):
4 plastic or paper cups

Measuring devices (per group of four):
graduated cylinders or measuring cups, optional

Background Information
An adaptation is any characteristic that helps an organism survive. The adaptation may be in the organism's physical appearance, the way its body functions, or the way it behaves. These changes occur through interaction with living and nonliving things in the environment.

An insect's mouthparts are a set of structures used for eating. They surround the insect's mouth. There are two basic types of insect mouths and mouthparts—those adapted for chewing and those adapted for sucking. Many insects have variations or combinations of the two basic types of mouths. For example, the mosquito has a piercing-sucking mouth, a housefly has a sponging mouth, and wasps and bees have chewing-lapping mouths.

Chewing insects have two grinding jaws called mandibles. They move sideways and are lined with teeth in most species. The jaws are also used for cutting or tearing off food. They have a second pair of less powerful jaws, called maxillae, behind the mandibles that are used to push the food down the throat. They also have two lips or flaps that hang down over the mouthparts and cover the front of the mouth. The upper lip is called the labrum and the lower lip, the labium. Some examples of chewing insects are grasshoppers, crickets, beetles, cockroaches, and termites.

Sucking insects have mouthparts adapted from the basic chewing mouth structure to suit their feeding habits. The labium in some insects has become a long, grooved beak with four slender, sharp needles called stylets. Stylets are used for piercing and then sucking up juices or blood. In butterflies and moths, the mandibles have lengthened forming a long drinking tube called a proboscis. This tube coils up when the insect is not using it to gather liquids. The mandibles in horseflies have become curved swords that can slash an animal's skin and its maxillae have developed into sharp-pointed rods that can penetrate and extract fluids from the skin.

Management

1. This activity can be done in one of two ways. The first student page has students simulating the different types of insect mouthparts and how they work. It is a good activity to do by itself or as an introductory lesson for the second student page. The second student page allows students to do some measuring and graphing. The setup for each lesson is identical except for the inclusion of measuring cups or graduated cylinders in the second activity to quantify the amount of food collected by the mouthparts.
2. This activity is designed to be done in groups of four.
3. Prepare the materials for each group ahead of time. Each group will need the materials listed.
4. To make management easier, each student should pick one food source and collecting cup to keep for the entire activity. Each student will then use each of the four mouthparts, in turn, to try and collect the "food." It will become obvious that not all the mouthparts work well with all the food sources. For example, the chewing mouth (clothespin) will not work well on anything but the bits of paper, while the straws will not be able to collect the bits of paper. It is important for the students to discuss their experiences with each of the four mouthparts and come to a group consensus as to which mouth is adapted best to each food supply.

Procedure

1. Discuss differences between insects. Focus on methods of eating, different types of mouthparts, and various types of food sources.
2. Talk about how different insects have adapted or changed over time to meet environmental changes, food sources, etc.
3. Have students share the different types of mouthparts on various insects that they have observed.
4. Hand out the first student page. Show each type of mouth and discuss how it works. Show the students the four kinds of food sources.
5. Using a cup of water, demonstrate how to capture liquid in a straw with your finger. Note: Caution students not to use their mouths to suck up liquid in the straw. To capture liquid in the straw, lower it into the liquid, and place your finger on top of the straw, trapping the liquid inside. To release the liquid, lower the straw into a "collecting cup," and remove your finger from the top of straw, releasing the liquid. Explain that the piercing-sucking mouthpart (the pointed straw) is the only one that should be used to break through the paper towel or plastic wrap on the covered cup.
6. Hand out a small collecting cup to each student to collect the food gathered by each mouthpart. (All students will use all four mouthparts during the lesson, but will keep the same collecting cup and food supply.)
7. Discuss the activity sheet, and have students fill in their predictions as to which mouth is best suited to each food supply. Explain the mechanics of the lesson. Each "feeding period" should be about two minutes.
8. After students have tried each of the four mouthparts to collect their food, have them discuss their observations and come to a group consensus as to which mouth is best adapted to each food source. They should then fill in the rest of the student page from their observations.
9. The activity can be repeated at a later time using the second student page and having the students count and measure the amounts of each food that the mouthparts collect.

Connecting Learning

1. How does an insect's mouth affect its choice of food?
2. What would happen if all insects had the same type of mouthparts?
3. Where would you look for an insect that had a "sucking" type of mouth? ...chewing? ...piercing/sucking? ...lapping?
4. What are you wondering now?

Extensions

1. Make a list of insects that have each type of mouthpart. Which is the most common type?
2. Discuss how mouthparts relate to where an insect lives.

3. Look up insects in various reference materials and draw and label the different types of mouthparts.
4. Design insect mouthparts that would be good at collecting a common food item, such as sugar or fruit.

Curriculum Correlation

Literature

Mound, Laurence. *Eyewitness Books: Insect.* DK Publishing. New York. 2007.

Wangberg, James K. *Do Bees Sneeze? And Other Questions Kids Ask About Insects.* Fulcrum Publishing. Golden, CO. 1997.

Science

What kinds of adaptations for eating do mammals have? What about reptiles, birds, amphibians, and fish?

Math

Test each mouthpart again allowing a feeding period that is twice as long. Did you eat twice as much?

Geography

Identify as many insects in your area as possible. Are there more insects with one particular kind of mouth? Identify the food source for each insect.

Table Manners

Key Question

How are insects adapted to eating certain types of food?

Learning Goals

Students will:

- simulate food gathering with four different types of insect mouths, and
- determine which mouth is most effective for different food sources.

Table Manners

Which mouthpart is best adapted to each food source?

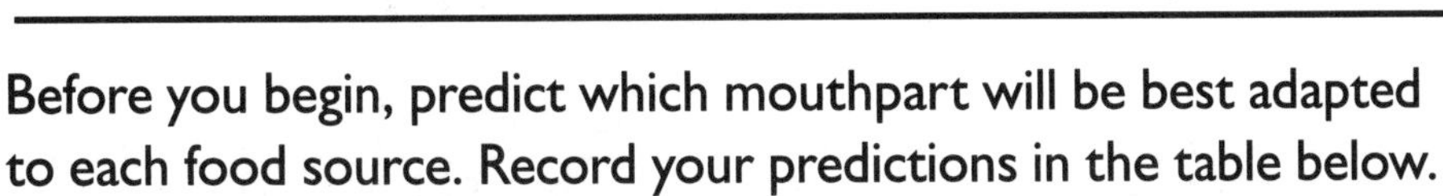

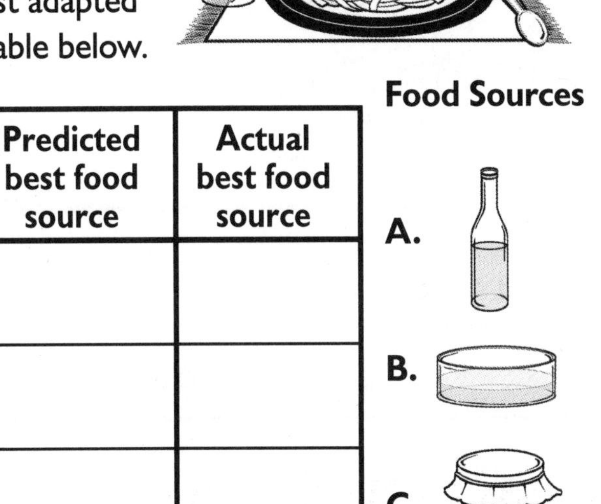

Before you begin, predict which mouthpart will be best adapted to each food source. Record your predictions in the table below.

Mouthpart	Predicted best food source	Actual best food source
Chewing		
Sucking		
Piercing/Sucking		
Sponging/Lapping		

Food Sources

A.

B.

C.

D.

Were your predictions correct? Why or why not?

Match the mouthpart with the critter.

How are these critters adapted to their food sources?

Table Manners

Which mouthpart will gather the most food?

Before you begin, predict which mouthpart will gather the most food. Record your prediction here.

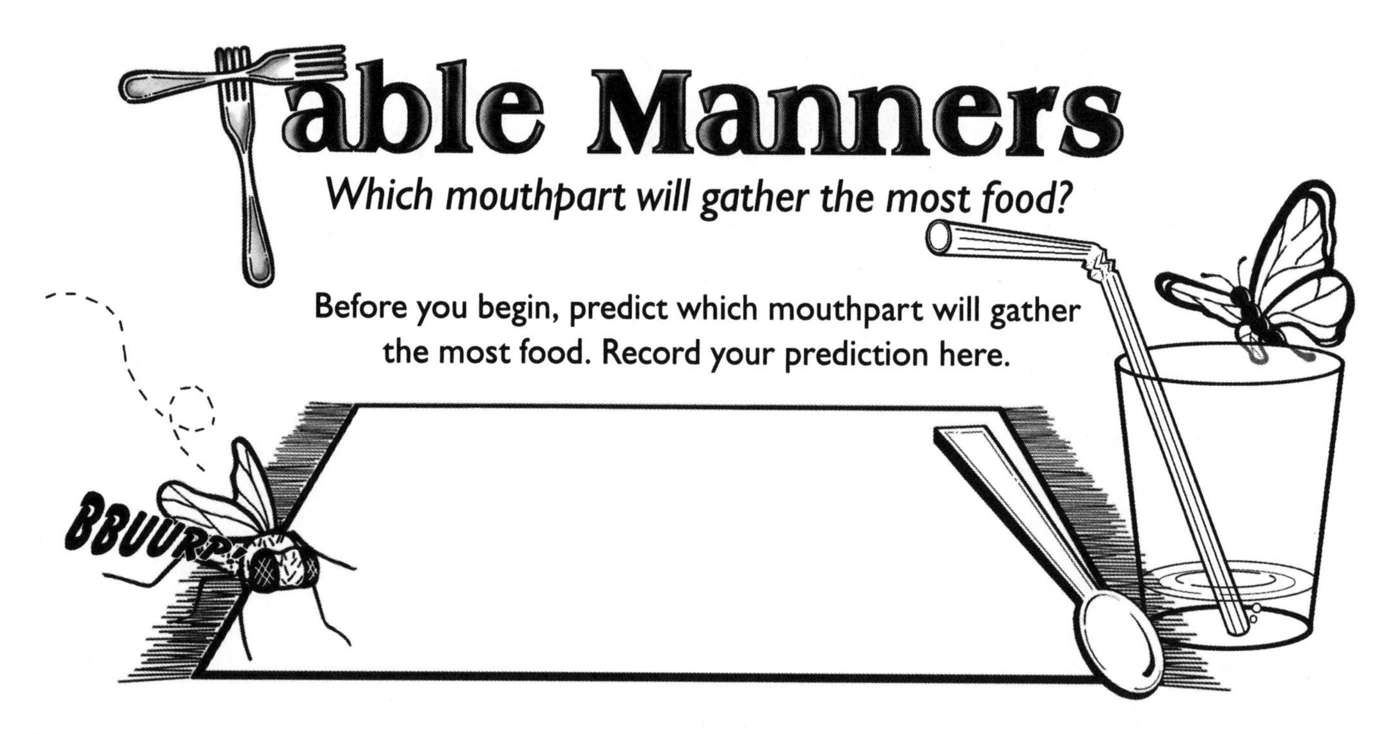

Record the amount of food eaten with each mouthpart at the end of the feeding session.

Mouthpart	Total food eaten
Chewing	pieces
Sucking	mL
Piercing/Sucking	mL
Sponging/Lapping	mL

1. Which mouthpart did you find the easiest to use? Why?

2. Which mouthpart was the most difficult to use? Why?

3. Which mouthpart is best suited to each kind of food source? Justify your response.

Table Manners

Make a bar graph showing how much food was collected using each of the mouthparts. Be sure to label the blank side of each graph with the appropriate numbers.

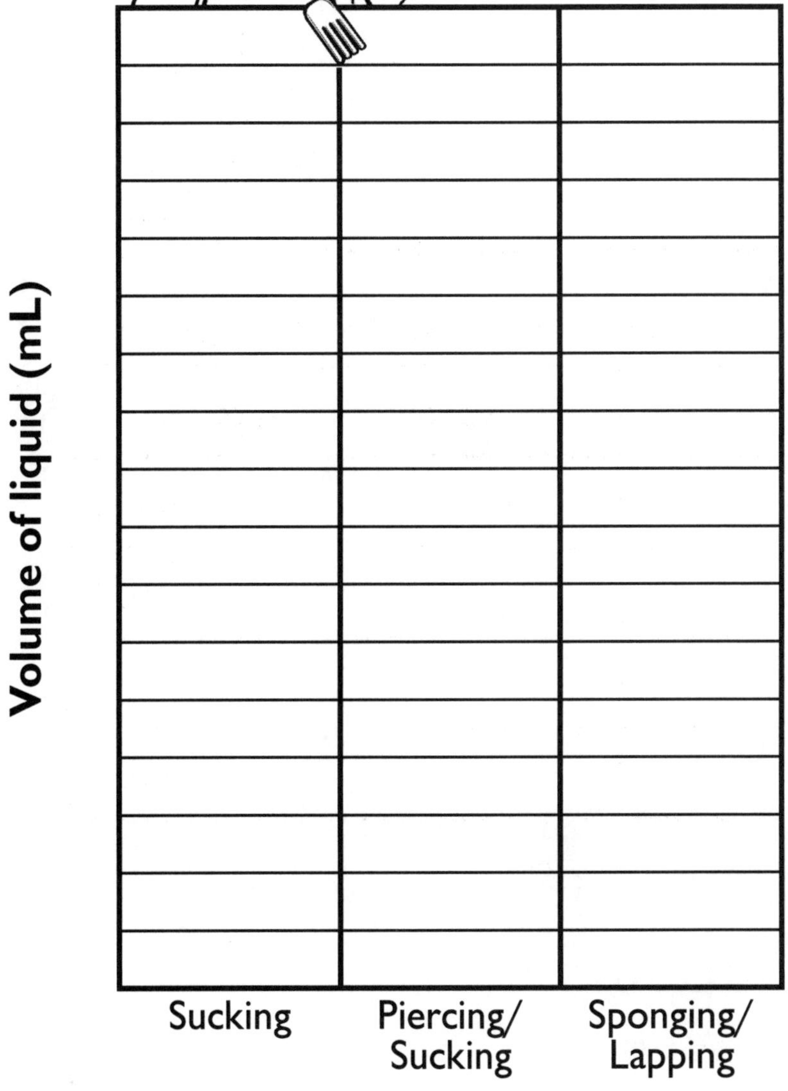

Volume of liquid (mL)

Sucking

Piercing/ Sucking

Sponging/ Lapping

Number of pieces

Chewing

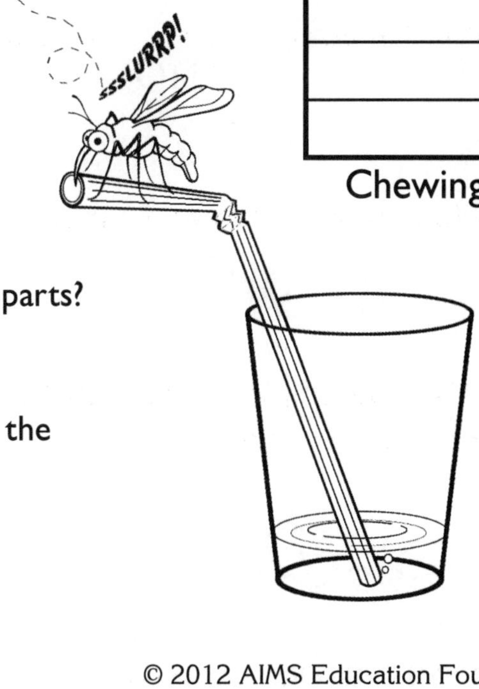

What do your graphs tell you about the different mouthparts?

Based on the graphs, which mouthpart would you say is the most effective? Why?

Table Manners

Connecting Learning

1. How does an insect's mouth affect its choice of food?

2. What would happen if all insects had the same type of mouthparts?

3. Where would you look for an insect that had a "sucking" type of mouth? ...chewing? ...piercing/sucking? ...lapping?

4. What are you wondering now?

I'm Stuck on You

Topic
Adaptations

Key Questions
1. How are the tongues of frogs, toads, and chameleons adapted to allow them to catch their food?
2. How many times can you successfully catch an insect in 10 trials?
3. Where can you catch the most insects—on the floor, on the wall, or when they are hanging in the air?

Learning Goals
Students will:
- learn about frog, toad, and chameleon tongues;
- make a "sticky tongue"; and
- simulate how some frogs, toads, and chameleons catch their food.

Guiding Document
Project 2061 Benchmarks
- *In doing science, it is often helpful to work with a team and to share findings with others. All team members should reach their own individual conclusions, however, about what the findings mean.*
- *Plants and animals have features that help them live in different environments.*
- *A model of something is different from the real thing but can be used to learn something about the real thing.*
- *Make something out of paper, cardboard, wood, plastic, metal, or existing objects that can actually be used to perform a task.*

Math
Number sense
Graphing

Science
Life science
    adaptations

Integrated Processes
Observing
Predicting
Comparing and contrasting
Generalizing
Applying

Materials
For the class:
chart paper
string
tape
pictures of insects
large paper tree, optional (see *Management 5)*
class graph (see *Management 8)*
Velcro™ strips or dots (see *Management 2)*
The Wide-Mouthed Frog (see *Curriculum Correlation)*
Internet access (see *Management 10)*

For each student:
1 party blower (see *Management 1)*
2 squares of Velcro™ (see *Management 2)*
1 paper insect
student pages

Background Information
The tongues of frogs, toads, and chameleons are wonderful examples of how adaptations make animals better able to survive in their environments. Many frogs, toads, and chameleons have long, sticky tongues that enable them to catch food. The tongues of frogs and toads are covered with slimy mucus to which their prey adheres. Their tongues can extend far out of their mouths to catch insects. Once they have pulled an insect into their mouths, they swallow their meal whole. A frog blinks as it swallows. Its eyeballs help force the food down.

A chameleon's tongue stays bunched up inside its mouth until it is time to catch a grasshopper or other insect. When the tongue shoots out, it is as long as the total length of the chameleon's body and tail. The insect sticks to the club-like padded tip. The tongue and insect are then reeled back into the mouth of the chameleon.

Management
1. Purchase the small-sized party blowers. Do not buy party blowers that make noises or have extra decorations on them that may distract many students. Plain party blowers work best.
2. The Velcro™ can be purchased in squares or in long strips by the yard. If you purchase the strips, cut one-inch squares prior to the activity. Purchase the Velcro™ that has a sticky backing to save time in preparing the "sticky tongues."

3. Designate three different areas of the room that will be used for hunting insects. One area will be on the floor (ground). Another area will be for insects that are hanging in midair (flying), and the third area will be for insects that are on a wall (tree).
4. Duplicate and cut out enough pictures of insects for each student to have one. Duplicate three extra sets of at least 10 each to use at the stations. You will need to stick the piece of fuzzy-sided Velcro™ onto the center of each insect for the extra sets.
5. To set up the first station, scatter one set of the grasshoppers over a small area on the floor for a group of two or four students to hunt ground insects. For the second station, glue or tape a set of flies to the ends of string and hang them from the ceiling. They should hang no higher than the eye level of students.

For the third station, tape a set of beetles to the wall to simulate insects that are on the sides of trees, bushes, buildings, etc. You can make a large paper tree to mount on the wall to lend an air of realism to this station.

6. Assign partners or groups of students to each area so that everyone is not at the same area at one time.
7. This activity is divided into three parts. The first part involves students in the exploration of party blowers and the construction of the "sticky tongues." The second part has paired students seeing how many insects they can catch in 10 trials. This part is only to provide students with practice in catching insects and recording and tabulating data. The third part has students rotating through stations to determine which is the easiest location for catching insects: on the ground (floor), in midair, or on a tree (wall).
8. Make a class graph on chart paper or bulletin board paper. Label the three columns: *On the Ground, In the Air,* and *On a Tree.*
9. Before making the sticky tongues, hand out the party blowers and allow children a few minutes to practice blowing. This free exploration is a must! Be sure that students are not too close to other students so that they avoid hitting each other in the eyes with the party blowers.
10. You will need a computer with Internet access and a projection system to show students the introductory video.

Procedure

Part One

1. Read *The Wide Mouth Frog* to the students and then lead them in a discussion about what frogs eat and how they might catch their food. Record their thoughts on large chart paper.
2. Take the discussion further with reference to reptiles such as chameleons and other amphibians such as toads by asking *Key Question 1:* How are the tongues of frogs, toads, and chameleons adapted to allow them to catch their food?
3. Show students the National Geographic video on chameleons (see *Internet Connections).* Discuss why frogs, toads, and chameleons need long, sticky tongues.
4. Tell the students that they are going to pretend to be one of these animals and will be making a sticky tongue that will allow them to catch an insect in a way similar to the animals they have been learning about.
5. Distribute the party blowers. Allow ample time for exploration. This may be a bit chaotic at first but well worth the time spent so you can get their attention for the real purpose of the activity!
6. Caution the students to keep a distance from their classmates to avoid hitting each other in the eyes.
7. Once the students have finished exploring their party blowers, lead them through a discussion about how these blowers are very much like the tongues of frogs, toads, and some lizards.

8. Distribute a square of the looped portion of the Velcro™ to each student.
9. Direct them to blow their party blowers so they are extended out as far as possible. They will need to keep them in this position.
10. Have a partner peel the back off the rough piece of Velcro™ to reveal its sticky surface and stick it on the **underside** of the end of the party blower.

11. Instruct the students to let the party blowers roll back up and put them aside. Have them help their partners affix the Velcro™ in the same way.
12. Allow each student to choose an insect and color it.
13. Distribute the companion pieces of Velcro™ (smooth/fuzzy), and show them how to affix the Velcro™ to the center of their insects.

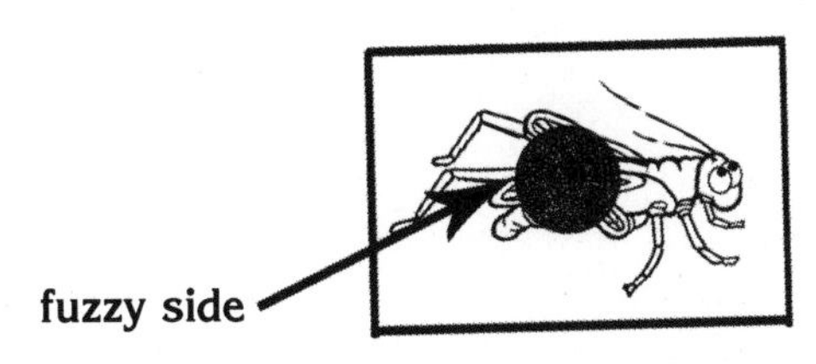

14. Direct students to place their insects on the floor. Have them blow out their "sticky tongues" and catch their insects.
15. After catching an insect, have students peel them off their "tongues" and continue practicing.

Part Two

1. Ask *Key Question 2:* How many times can you successfully catch an insect in 10 trials?
2. Have students work in pairs. Hand out one copy of the first student page to each pair of students. Tell them that one student (*the animal)* will try to catch the insect and the other student (*the scientist)* will keep a record of the trials on the student page.
3. Instruct the students who are the *animals* to guess how many times they can catch an insect in 10 trials and record that number on the first student page.
4. Ask the *animals* to place their insects on the table and make 10 attempts to catch the insects with their tongues.
5. Each time the *animals* attempt to catch an insect, have the *scientists* make a check mark in the appropriate column on the student page.
6. After all 10 trials, have the *scientists* count up and record the total number of successful catches.
7. Instruct the *animals* and *scientists* to switch roles and repeat this procedure.

Part Three

1. Ask *Key Question 3:* What is the most successful place to catch an insect—on the floor, on a wall, or hanging in the air?
2. Identify the three insect-collection areas already set up in the room.
3. Have students work with partners or small groups.
4. Give instructions for how you would like students to rotate through the insect-collection areas and distribute the second student page to each student.
5. Tell students that in each area they will take turns trying to catch the insects 10 different times. Each person must record his or her own data on the student page.
6. After all students have been through the three stations, distribute the *Pest Strips* bar graphs. On each half of the page, have each student color the insects on the graph to represent his or her successful catches in *Part Three.*
7. Once students have recorded their results, have them cut out the strips on the right half of the page to put on a large class graph.
8. Combine the individual results into a class graph to see in which area the students were most successful.
9. Close with a final time of class discussion and sharing.

Connecting Learning

1. How does the tongue of a frog, toad, or chameleon help it survive?
2. What techniques did you use to catch the insects? [Answers will vary but may include: a fast flick of the tongue, a slow flick of the tongue, being close to the insect, or standing as far from the insect as possible.]
3. Which technique worked the best for you? Why do you say that?
4. Was anyone successful in catching an insect 10 times in a row?
5. Do you think frogs, toads, and chameleons catch their meals every time they try?
6. In which hunting place was it easiest for you to catch your insects? Why?
7. Where do you think it is easiest for frogs, toads, and chameleons to catch their food? Why?
8. Was your actual catch in *Part Two* close to what you had guessed? Why or why not?
9. Do you think your easiest catch was the easiest for the rest of the class? Explain.
10. What are you wondering now?

Extensions

1. Let students select an animal (frog, toad, or chameleon) and design a mask with a hole where the tongue would be. They will need to make the hole big enough for the party blower to fit through. The students will also need to cut two holes for eyes so they can find their insects!
2. Discuss the roles of predator and prey in this activity. [Frog is the predator; insect is the prey.] Then ask the students when an insect can be a predator and when a frog can be the prey. Have them name animals that would be predators of frogs. [birds, snakes, etc.]
3. Have students design a possible food chain that would include the insect and either a frog, toad, or chameleon. For example: Raccoon ← fish ← frog ← insect ← leaves.

Internet Connections

National Geographic—Crafty Chameleon
http://video.nationalgeographic.com/video/player/nat-geo-wild/shows-1/worlds-deadliest/ngc-crafty-chameleon.html
This two-minute video describes many of the chameleon's adaptations, including its amazing sticky tongue. Note: begins with an advertisement.

Curriculum Correlation

Moffett, Mark W. *Face To Face With Frogs.* National Geographic. Wasington, DC. 2010.

Clarke, Barry. *Eyewitness Books: Amphibian.* DK Publishing. New York. 2005.

Gibbons, Gail. *Frogs.* Holiday House. New York. 1994.

Schneider, Rex. *The Wide-Mouthed Frog.* Stemmer House Publishers. Gilsum, NH. 1994.

I'm Stuck on You

Key Questions

1. How are the tongues of frogs, toads, and chameleons adapted to allow them to catch their food?
2. How many times can you successfully catch an insect in 10 trials?
3. Where can you catch the most insects—on the floor, on the wall, or when they are hanging in the air?

Learning Goals

Students will:

- learn about frog, toad, and chameleon tongues;
- make a "sticky tongue"; and
- simulate how some frogs, toads, and chameleons catch their food.

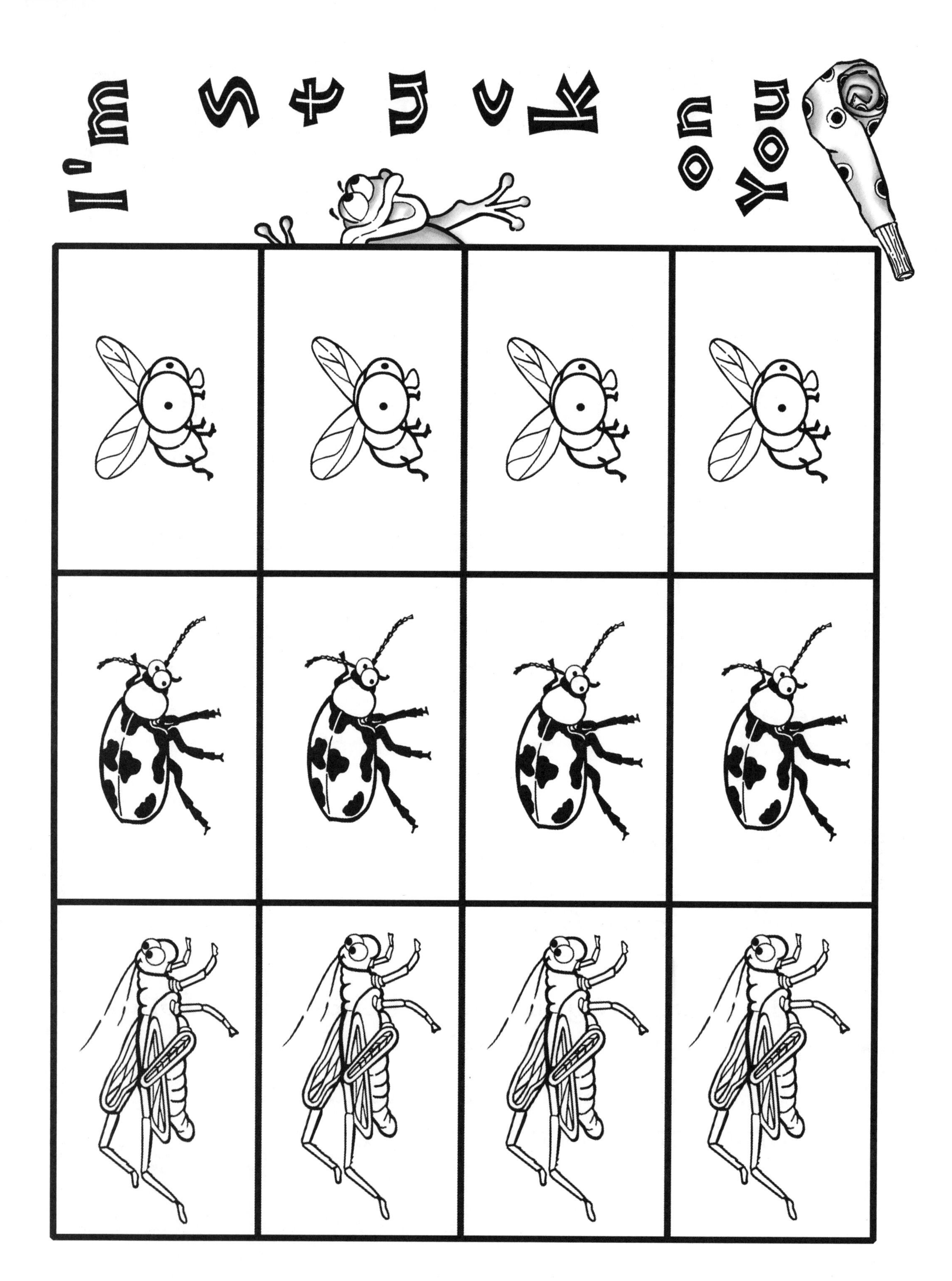
I'm Stuck on You

I'm Stuck on You

Animal ________ Scientist ________

I think I will catch _____ insects.

Trial	Caught	Missed
1		
2		
3		
4		
5		
6		
7		
8		
9		
10		

I actually caught _____ insects.

Animal ________ Scientist ________

I think I will catch _____ insects.

Trial	Caught	Missed
1		
2		
3		
4		
5		
6		
7		
8		
9		
10		

I actually caught _____ insects.

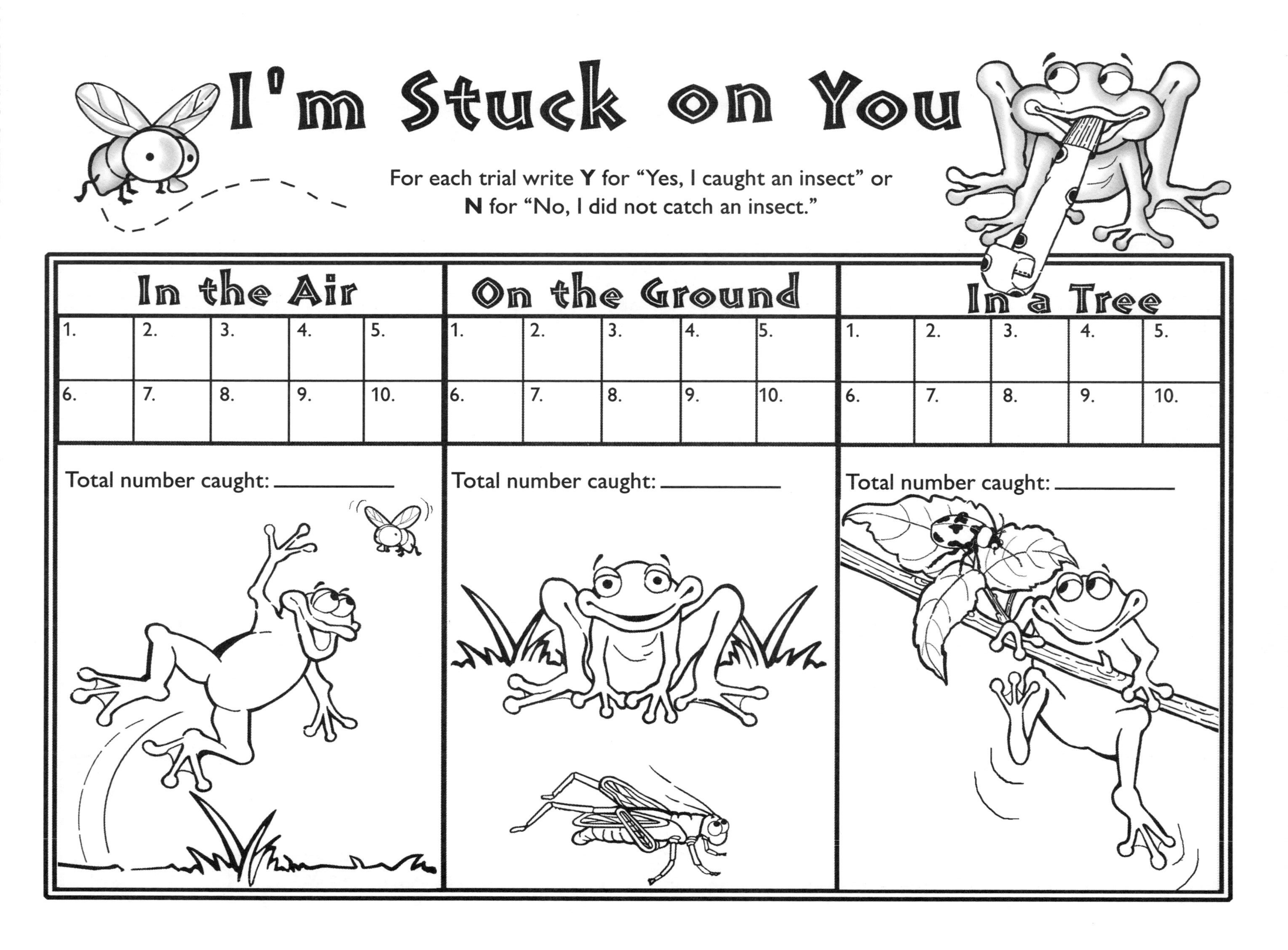

I'm Stuck on You

For each trial write **Y** for "Yes, I caught an insect" or **N** for "No, I did not catch an insect."

In the Air					On the Ground					In a Tree				
1.	2.	3.	4.	5.	1.	2.	3.	4.	5.	1.	2.	3.	4.	5.
6.	7.	8.	9.	10.	6.	7.	8.	9.	10.	6.	7.	8.	9.	10.

In the Air — Total number caught: ___________

On the Ground — Total number caught: ___________

In a Tree — Total number caught: ___________

Pest Strips

Record the number of times you caught each bug by coloring in that many squares. Color both sets of strips.

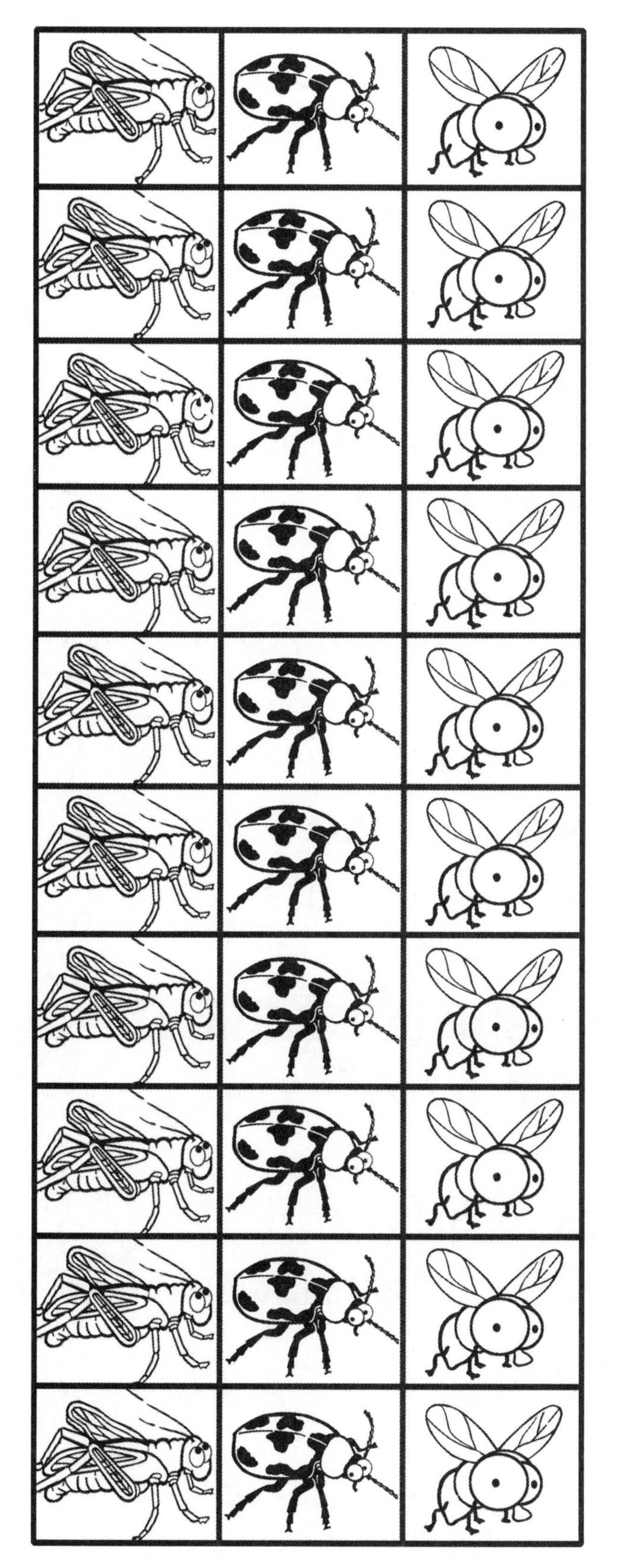

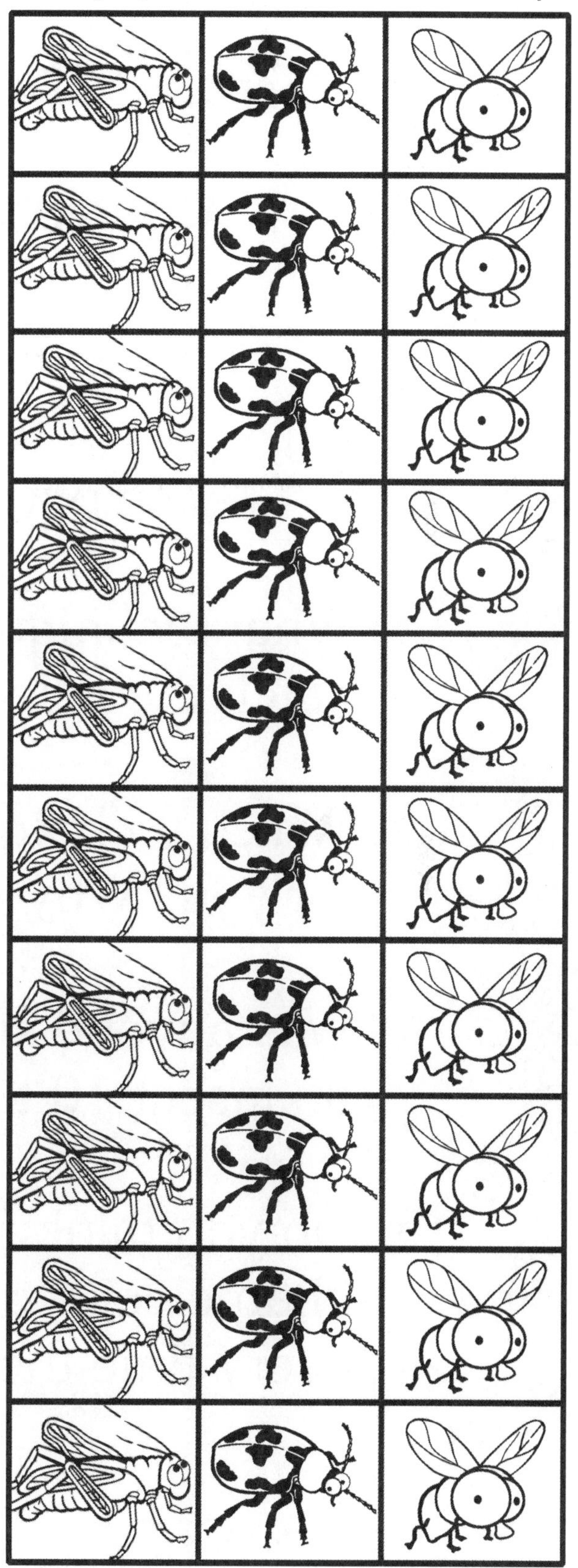

I'm Stuck on You

Connecting Learning

1. How does the tongue of a frog, toad, or chameleon help it survive?

2. What techniques did you use to catch the insects?

3. Which technique worked the best for you? Why do you say that?

4. Was anyone successful in catching an insect 10 times in a row?

5. Do you think frogs, toads, and chameleons catch their meals every time they try?

I'm Stuck on You

CONNECTING LEARNING

Connecting Learning

6. In which hunting place was it easiest for you to catch your insects? Why?

7. Where do you think it is easiest for frogs, toads, and chameleons to catch their food? Why?

8. Was your actual catch in *Part Two* close to what you had guessed? Why or why not?

9. Do you think your easiest catch was the easiest for the rest of the class? Explain.

10. What are you wondering now?

Topic
Adaptations: webbed feet

Key Questions
How does the webbing in frogs' feet help them swim?

Learning Goals
Students will:
- learn about the variety in the structure of the feet of different types of frogs,
- make paddle boats that simulate two different kinds of frog feet,
- use webbed and non-webbed hands to compare motion through water, and
- generalize what kind of foot is best for frogs that live in the water.

Guiding Documents
Project 2061 Benchmarks
- *Some animals and plants are alike in the way they look and in the things they do, and others are very different from one another.*
- *Plants and animals have features that help them live in different environments.*
- *Different plants and animals have external features that help them thrive in different kinds of places.*

NRC Standard
- *Each plant or animal has different structures that serve different functions in growth, survival, and reproduction. For example, humans have distinct body structures for walking, holding, seeing, and talking.*

Math
Measurement
linear

Science
Life science
adaptations
frog feet

Integrated Processes
Observing
Predicting
Collecting and recording data
Comparing and contrasting
Relating
Applying
Generalizing

Materials
For each pair of students:
2 or 3 Styrofoam meat trays (see *Management 1)*
2 or 3 rubber bands (see *Management 3)*
2 large paper clips (see *Management 5)*
ruler
one pair of clear plastic gloves (see *Management 9)*
boat pattern page
journal page, one per student
scissors

For the class:
several tubs of water (see *Management 4)*
clear plastic wrap (see *Management 8)*
clear packaging tape

Optional:
non-fiction books and pictures of frogs showing their feet
Internet access (see *Internet Connections)*

Background Information
An adaptation is a characteristic of an animal or plant that enables it to live successfully in a particular place. Adaptations may be physical, or they may be part of an animal's behavior. The animals and plants that are best adapted to their environments are most likely to survive and reproduce, passing along their adaptive characteristics to their offspring.

Frogs all have four toes on their front feet and five toes on their back feet. Beyond that, the design of the feet varies greatly, providing excellent examples of adaptations. A close look at the legs and feet of a particular type of frog will tell a great deal about how it lives, its behavior, and the habitat in which it lives. Frogs that live mostly on land usually have shorter legs, which are better suited for walking and climbing. Some of these frogs have broad feet with short, stubby toes, ideal for burrowing into the ground. Tree-climbing frogs have large, round, sticky pads on their toes that work like suction cups to help them cling to branches, leaves, and other often slippery surfaces. Water-dwelling frogs have comparatively long, strong legs with webbed feet that help them swim better and faster.

Webbing of feet is an adaptation shared by ducks, geese, swans, otters, salamanders, frogs, and certain other animals that spend a significant part of their lives in water. In this activity, students will experience the advantage of having webbed feet as an animal moves through water. In *Part One*, they will make paddle boats with two different kinds of paddles representing webbed and non-webbed feet. In *Part Two*, they will experience the effects of webbing as they move their own gloved hands, one of which is webbed, through a container of water.

Management

1. You can get Styrofoam meat trays at a grocery store or use Styrofoam take-out boxes. Styrofoam plates should be used only if they are sturdy; many are too flimsy and will buckle under the pressure of the rubber band.
2. Depending upon the cutting skills of your students, you may need to cut the paddles ahead of time or assist students in doing so themselves.
3. Rubber bands should be medium in size and fairly thin. You may need to test several sizes to find the ones that work best.
4. If available, a small wading pool works very well. Otherwise, use individual tubs at least four inches deep and 24 inches long (longer is better).
5. Use a large paper clip on the front of the boat to keep it from flipping over.
6. Several non-fiction frog books and/or pictures should be available for looking closely at the structure of frogs, especially their feet. See also the websites listed.
7. Before doing *Part Two*, cut the clear plastic wrap into 6-inch by 3-inch strips—one for each student.
8. Inexpensive disposable clear plastic gloves used for handling food, etc., may be available from the school cafeteria or purchased at discount stores.

Procedure

Part One

1. Use pictures, books, or images from the Internet to show some of the variety in the structure of the feet of different types of frogs. Ask the students to offer suggestions for the purposes of the different feet they see. Call the students' attention to the differences between front and back feet as well. (See *Background Information*.)
2. Tell the students that they will be making boats that are propelled in a way similar to how the frogs' feet propel them through the water. The boats will have two different paddles representing two of the different types of frog feet.
3. Have students get into pairs and distribute the materials. Go over the construction instructions as a class. Assist students as necessary during boat construction.
4. Demonstrate the method of sailing the boat with the webbed paddle:

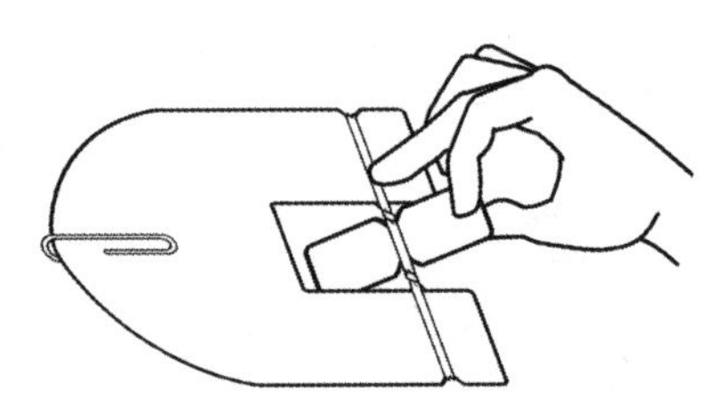

   (a) Center the paddle in the rubber band.
   (b) Rotate the paddle from the top toward yourself.
   (c) Use five twists of the paddle for consistency.
   (d) Use one hand to launch the boat. It helps to push down very slightly so that the front of the boat is just a little bit above the surface of the water.
5. Have the students launch the web-paddled boat and observe. Discuss the results, including distance traveled, speed, course, and anything else they notice. Allow time for the students to troubleshoot and fix problems. If the boats do not go in a relatively straight line, the paddle may be hitting the side of the slot. If the boat goes backwards, the paddle was wound in the wrong direction. If it flips, another paper clip may need to be added to the front, or the students may be pushing down too hard on the back of the boat. If necessary, launch the boats again until the students are successful.
6. Ask the students to predict what might happen using the non-webbed paddle. Repeat *Procedure 5* using that paddle. Compare the results, discuss, and record on the journal page.
7. Launch both boats side by side, comparing the speed and distance traveled. Encourage the students to do several test launches until their results are consistent.
8. Bring the class together around one tub. Have one team launch two boats side by side. Ask the students to reflect upon which boat traveled faster and farther, and what difference was made by each type of paddle.
9. Have students look at the pictures of the frogs' feet and ask the students which paddle looks most like which kind of foot. Talk about what the advantages of having webbed feet might be for a frog that lives primarily in the water.

Part Two

1. Remind the students of what they observed in *Part One*. Explain that in this part of the activity, they will experience firsthand what it feels like to have webbed feet (hands).
2. Distribute two pairs of gloves to each team. Have one student in each team put on a pair of gloves. Have him or her spread the fingers on one hand as far apart as possible. Direct the other student to cover the four fingers with plastic wrap and tape it securely, forming webbing.

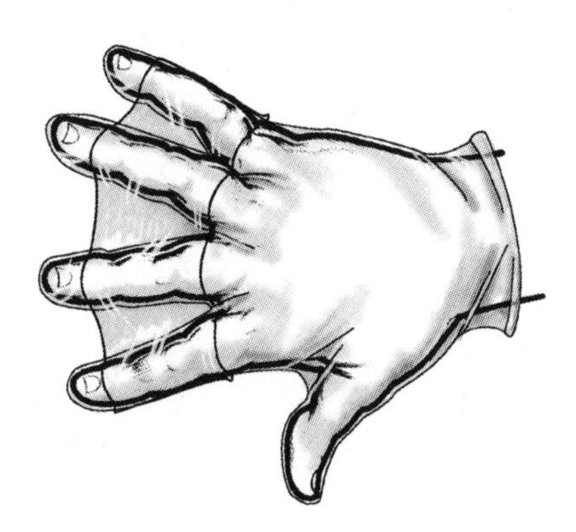

3. Now tell the student wearing the gloves to drag both hands through the tub of water with fingers on both hands spread wide apart. Ask him or her to compare what he/she feels with each hand, webbed and non-webbed, while the other student observes and describes the motion of the water.
4. Instruct the partners to switch places and repeat the procedure, then record their observations on the journal page.
5. Bring the class together again around a tub and choose one team to demonstrate. Ask the students to share their observations. Discuss the advantage of having webbed feet for frogs and certain other water-dwelling animals. Encourage students who have used swim fins to relate that experience with what they have done in this activity.

Connecting Learning

1. Looking at the pictures, how are the frogs' feet different? How would each of these types of feet help the frog to live in its habitat?
2. How are the boat paddles similar to different kinds of frog feet?
3. Which kind of paddle moved the boat farther and the faster? Why? Did the other teams have the same results you did? Why or why not?
4. Describe how your webbed hand felt when it moved through the water. ...your non-webbed hand. What difference did the webbing make? What did it do to the water?
5. How did the boats' paddles help them move? How do a frog's feet help it swim?
6. If you were to design a foot for a water-dwelling animal, what would it look like? Explain your thinking.
7. Which would make a better paddle for a boat—a fork or a spoon? Explain.
8. What are you wondering now?

Extensions

1. Find a toy rubber ball covered with suction cups. Use it to illustrate the foot adaptations of climbing frogs.
2. Research other animals with webbed feet.
3. Encourage students to improve upon the design of the boat paddle.

Internet Connections

Exploratorium Frogs
http://www.exploratorium.edu/frogs/mainstory/index.html
This six-page story describes adaptations of frogs. Page two discusses frogs' feet and has photographs.

Frogland
http://allaboutfrogs.org/froglnd.shtml
Under the "WEIRD Frog Facts" link is a page on frog feet. Lots of other frog information is also available.

Curriculum Correlation

Clarke, Barry. *Eyewitness Books: Amphibian*. DK Publishing. New York. 2000.

Gibbons, Gail. *Frogs*. Holiday House. New York. 1994.

Moffett, Mark W. *Face to Face With Frogs*. National Geographic. Washington, DC. 2010.

Home Link

Encourage students to take their boats home to demonstrate and continue using.

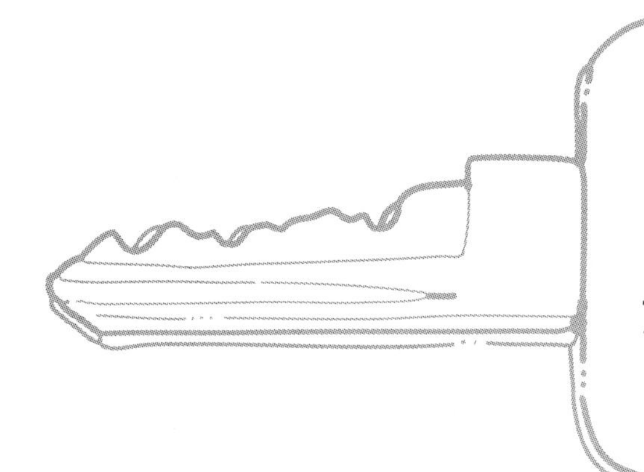

Key Question

How does the webbing in a frog's feet help it swim?

Learning Goals

Students will:

- learn about the variety in the structure of the feet of different types of frogs,
- make paddle boats that simulate two different kinds of frog feet,
- use webbed and non-webbed hands to compare motion through water, and
- generalize what kind of foot is best for frogs that live in the water.

Wonderful Webbed Feet

Making the Boat Pieces

- Carefully cut out each pattern.
- Trace the patterns onto the Styrofoam plates or trays.
- Trace two boat patterns.
- Cut out the two boats and the two paddles.

Assembling the Boats

- Put a large paper clip on the nose of each boat.
- Put a rubber band around each boat so that it fits in the notches.
- Slide one of the paddles into the cut out section of one boat. Put it between the rubber bands so that the notches line up.
- Twist the paddle toward you five times.
- Carefully put the boat into the water and let go of the paddle.
- Repeat with the other paddle and the other boat.

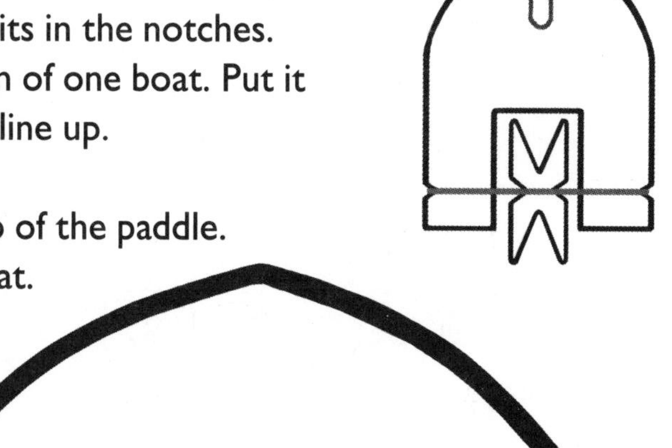

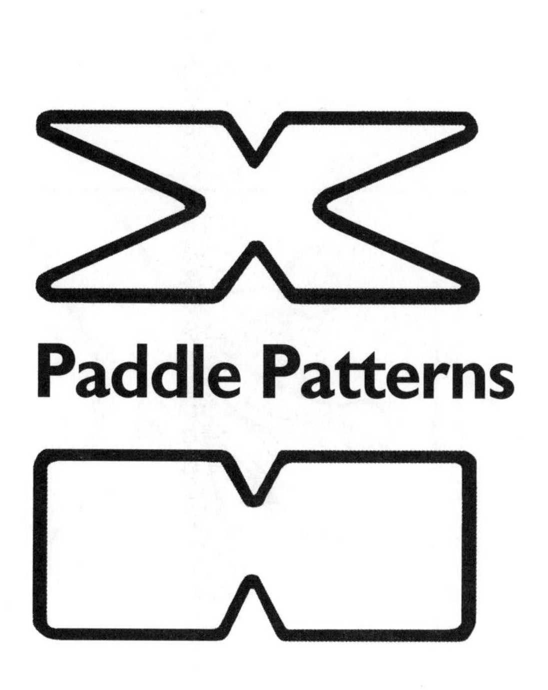

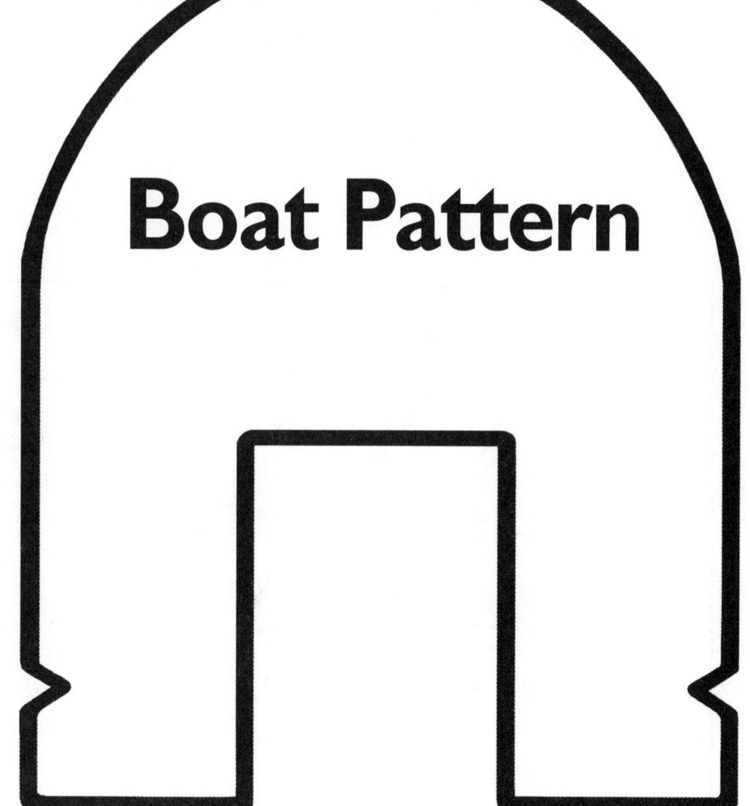

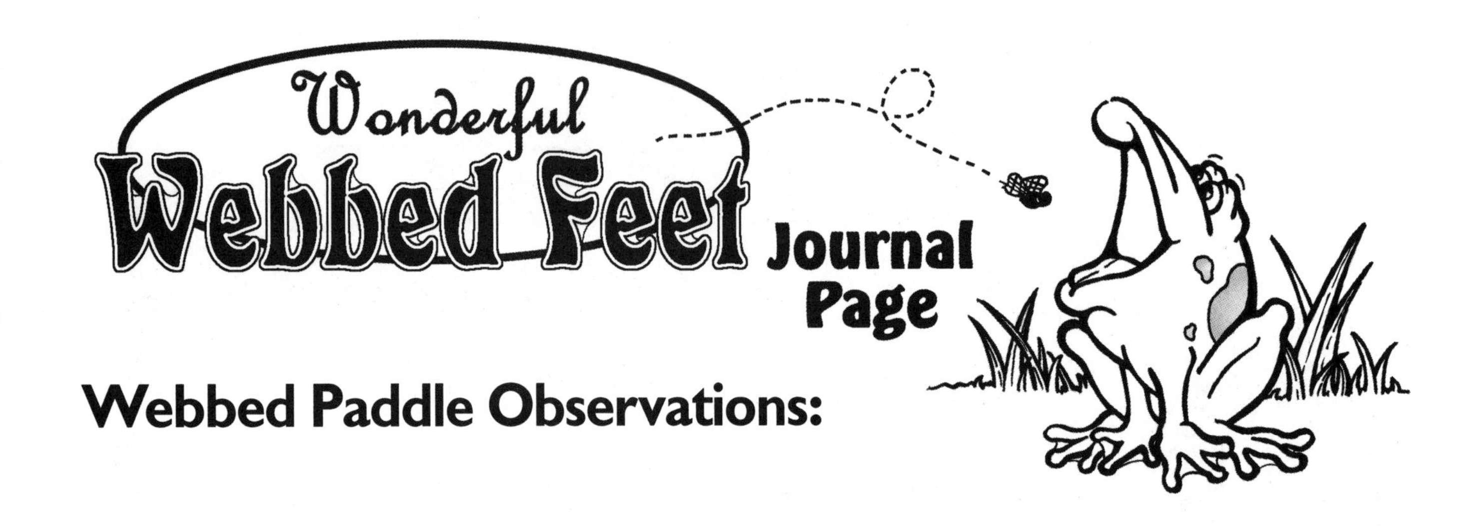

Webbed Paddle Observations:

Non-webbed Paddle Observations:

Glove Observations:

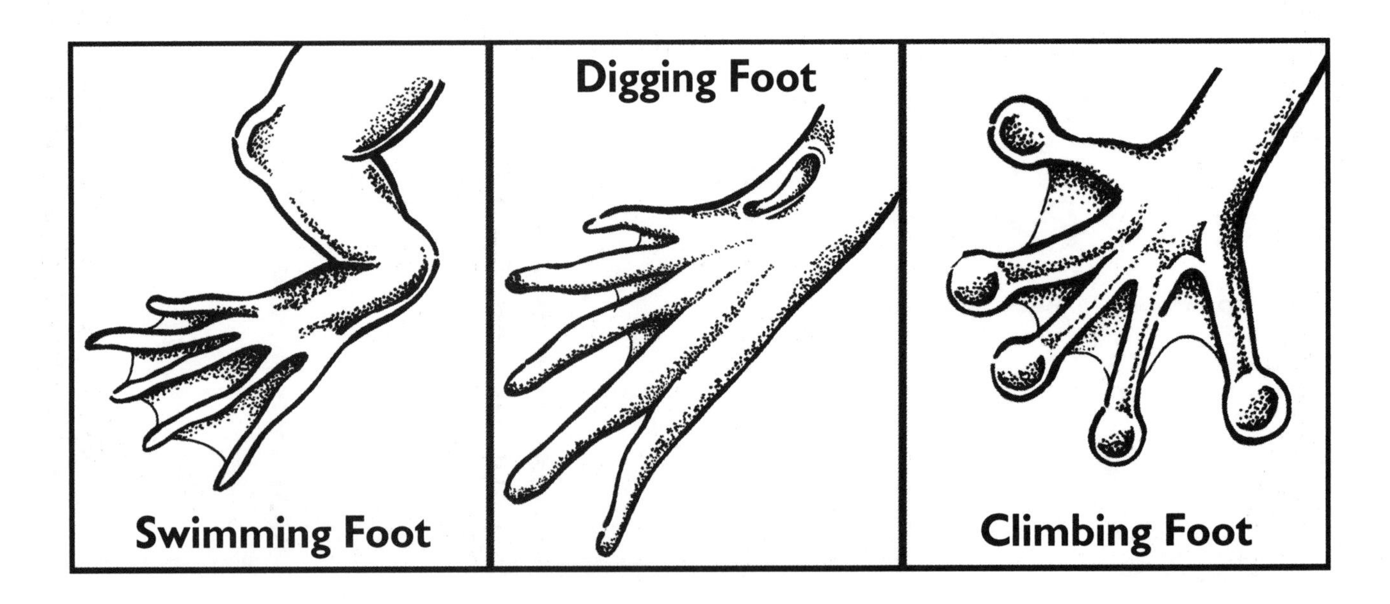

CONNECTING LEARNING

Connecting Learning

1. Looking at the pictures, how are the frogs' feet different? How would each of these types of feet help the frog to live in its habitat?

2. How are the boat paddles similar to different kinds of frog feet?

3. Which kind of paddle moved the boat farther and the faster? Why? Did the other teams have the same results you did? Why or why not?

4. Describe how your webbed hand felt when it moved through the water. ...your non-webbed hand. What difference did the webbing make? What did it do to the water?

CONNECTING LEARNING

Connecting Learning

5. How did the boats' paddles help them move? How do a frog's feet help it swim?

6. If you were to design a foot for a water-dwelling animal, what would it look like? Explain your thinking.

7. Which would make a better paddle for a boat—a fork or a spoon? Explain.

8. What are you wondering now?

Bird Beaks and Fowl Feet

Topic
Adaptations

Key Question
How are the beaks and feet of different birds adapted to their needs?

Learning Goals
Students will:
- match descriptions of different bird beaks and feet to illustrations,
- learn the purposes of these adaptive characteristics, and
- create imaginary birds that match to characteristics given by the teacher.

Guiding Documents
Project 2061 Benchmark
- *Individuals of the same kind differ in their characteristics, and sometimes the differences give individuals an advantage in surviving and reproducing.*

NRC Standard
- *Each plant or animal has different structures that serve different functions in growth, survival, and reproduction. For example, humans have distinct body structures for walking, holding, seeing, and talking.*

Science
Life science
  adaptations

Integrated Processes
Observing
Comparing and contrasting
Analyzing
Recording
Applying

Materials
Adaptations for the Birds rubber band book
#19 rubber bands
Beak and feet cards, one set per student
Scissors
Glue sticks
Stapler
Build a Bird pages

Background Information
Organisms can only survive in environments where their needs can be met. Over time, plants and animals may change, or adapt, to be better suited to surviving in their surroundings. These changes are known as adaptations. These adaptations can be everything from protective coloration to complex body systems that allow an animal to live in harsh conditions.

Birds are a wonderful example of many different adaptive characteristics. The beaks and feet of birds are uniquely adapted to their lifestyles, and much can be told about a bird by simply observing these two features. This activity allows students to learn some of the basic feet and beak types of birds and then apply what they have learned to make imaginary birds with specific characteristics.

Management
1. The second part of the activity, which uses the *Build a Bird* pages, can be done in groups, or each student can have his or her own copy of the pages.
2. To assemble the *Build a Bird* book, align the pages and staple along the left edge. Cut along the solid line through the center, being careful not to cut all the way across the pages. Fold the top and bottom sections back one at a time to reveal the different possible combinations.

Procedure
1. Distribute the rubber band book, beak and feet cards, scissors, and a glue stick to each student. Have students assemble the rubber band book and cut apart the cards.
2. Allow time for students to read the book and glue the pictures in the appropriate spaces.
3. Discuss each kind of beak and foot, and have students share which picture they put with each description. If there are differences, identify the correct response.
4. Tell students that now that they know some of the purposes for different bird beaks and feet, you are going to have them create imaginary birds that match your descriptions.
5. Distribute the *Build a Bird* pages to each student or group. Show them how to align the pages, staple along the left edge, and cut along the solid line through the center of the pages. Be sure that they do not cut all the way to the stapled edge of the pages.

6. Describe the characteristics of the imaginary bird you would like students to create. For example, one that eats small rodents and lives in the water.
7. Have students/groups flip the pages of their books until they have come up with the beak/foot combination that matches the description you gave. Compare responses and discuss any differences.
8. Repeat as desired with other combinations of characteristics.

Connecting Learning

1. What are some of the different kinds of feet that birds can have? ...beaks?
2. What purposes do these different feet and beaks serve?
3. How did you decide which beaks and feet to use for the imaginary birds? Were there ever disagreements? Why or why not?
4. What would you guess that the feet of a sparrow would look like? [perching] Why?
5. What kind of beak do you think a teal has? [straining] Why?
6. What are you wondering now?

Extensions

1. Make a display showing all of the possible imaginary birds that can be made using the beak and feet combinations in the book. Have students create a name for each one and describe its characteristics.
2. Explore other adaptive characteristics of birds, such as body and/or appendage size as it relates to climate.

Solutions

Grasping

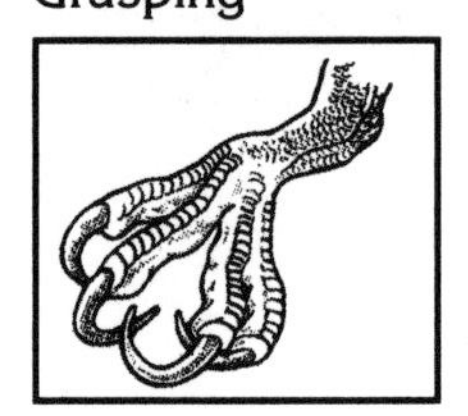

Cracking

Perching

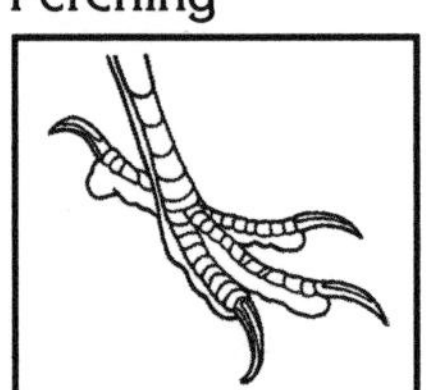

Tearing/Shredding

Swimming

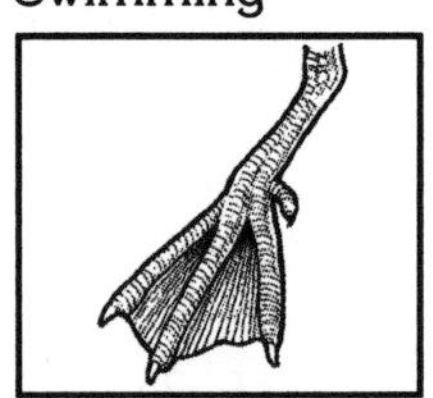

Spearing

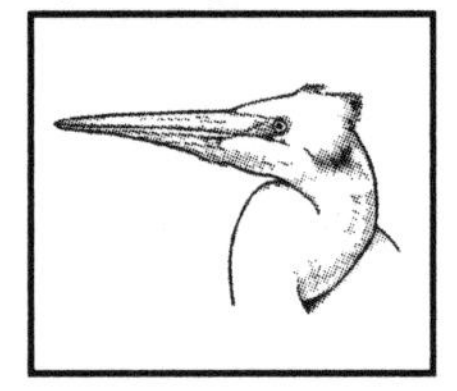

Scratching

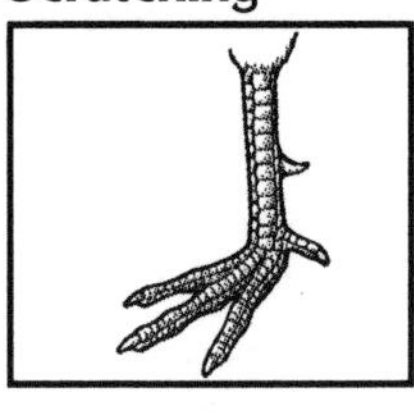

Probing/Slurping

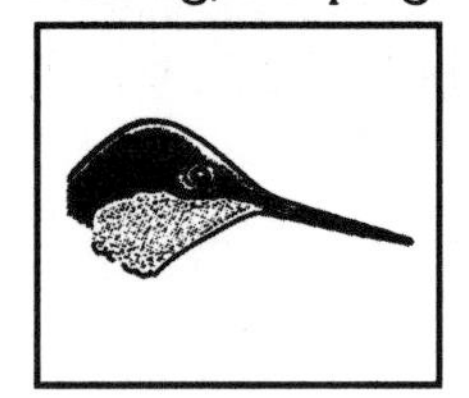

Wading

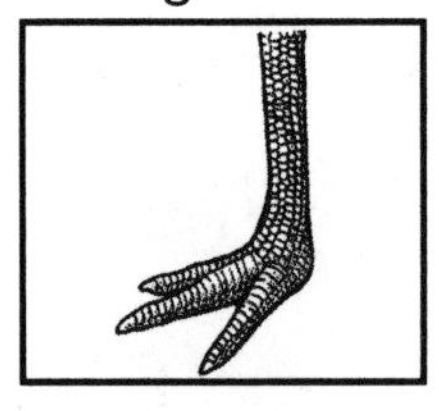

Straining

Climbing

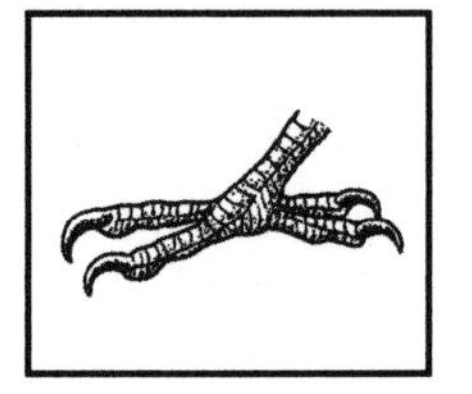

Chiseling

Bird Beaks and Fowl Feet

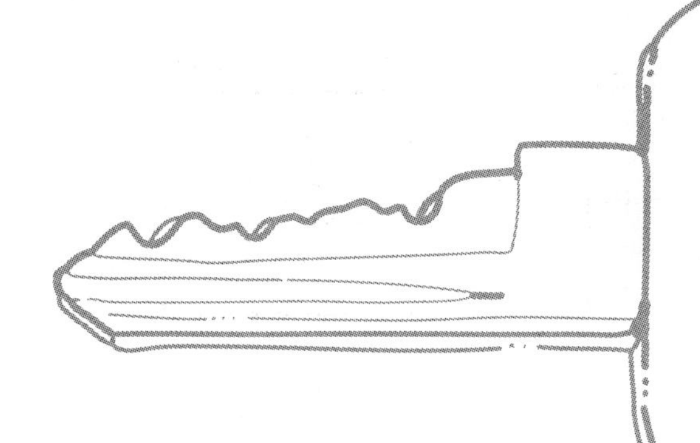

Key Question

How are the beaks and feet of different birds adapted to their needs?

Learning Goals

- match descriptions of different bird beaks and feet to illustrations,
- learn the purposes of these adaptive characteristics, and
- create imaginary birds that match to characteristics given by the teacher.

Adaptations are special characteristics that plants or animals have. These characteristics make it easier for them to survive. For example, penguins have an extra layer of fat and special feathers. This helps keep them warm and dry in the cold.

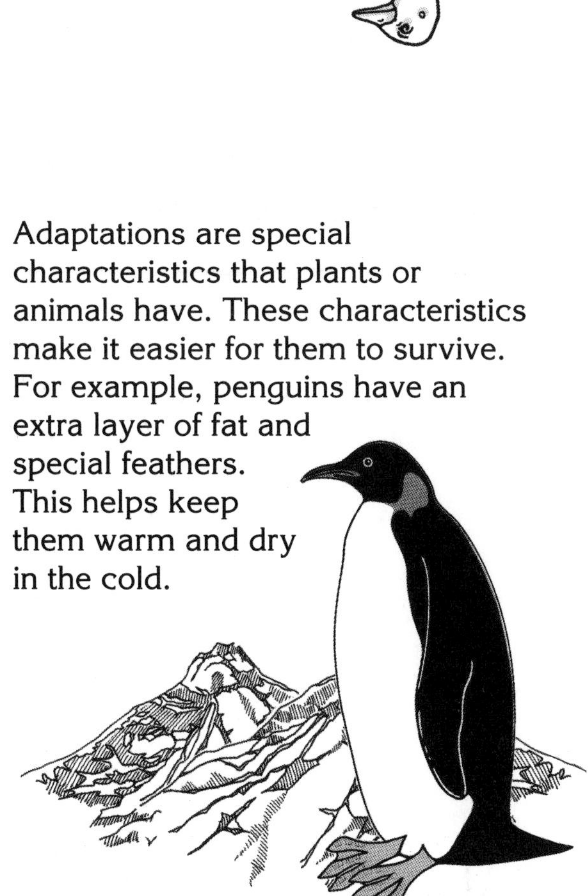

Chiseling

Long, thick chisel-like beaks allow the birds to chip away wood and tree bark.
Example: woodpeckers

Adaptations for the Birds

Probing/Slurping

Long, thin beaks allow the birds to get to the insides of flowers for nectar.
Example: hummingbirds

13

These feet have large curved talons to catch prey.
Examples: ospreys, hawks, eagles

4

Straining

Wide and flat or scoop shaped, these bills can strain out small plants and animals from water.
Examples: ducks, flamingoes

14

The beaks and feet of birds are adapted to what they eat and how they live. You will be reading about some of the different feet and beaks that birds have. Glue the picture that matches the description on each page.

3

11

Sharp, curved, and thick, these beaks allow the birds to tear meat.
Examples: hawks, owls, eagles

Tearing/Shredding

6

These feet are webbed and act like paddles in the water.
Examples: ducks, swans

Swimming

Spearing

Long, spear-like beaks are ideal for hunting fish.
Examples: herons, cranes

12

Perching

A long back toe allows birds with these feet to grab tightly to branches.
Examples: robins, jays, chickadees

5

Scratching

Three long front toes on these feet are used to scratch the soil to uncover seeds and insects. The back toe is short.
Examples: chickens, pheasants

7

Wading

The toes on these feet are long to make it easier to walk in the mud.
Examples: herons, cranes

8

Climbing

Two toes face forward and two toes face back, making it easier to climb the sides of trees or other vertical surfaces.
Examples: woodpeckers, cuckoos

9

Cracking

These beaks are short, thick, and conical.
They are ideal for eating seeds.
Examples: sparrows, cardinals, finches

10

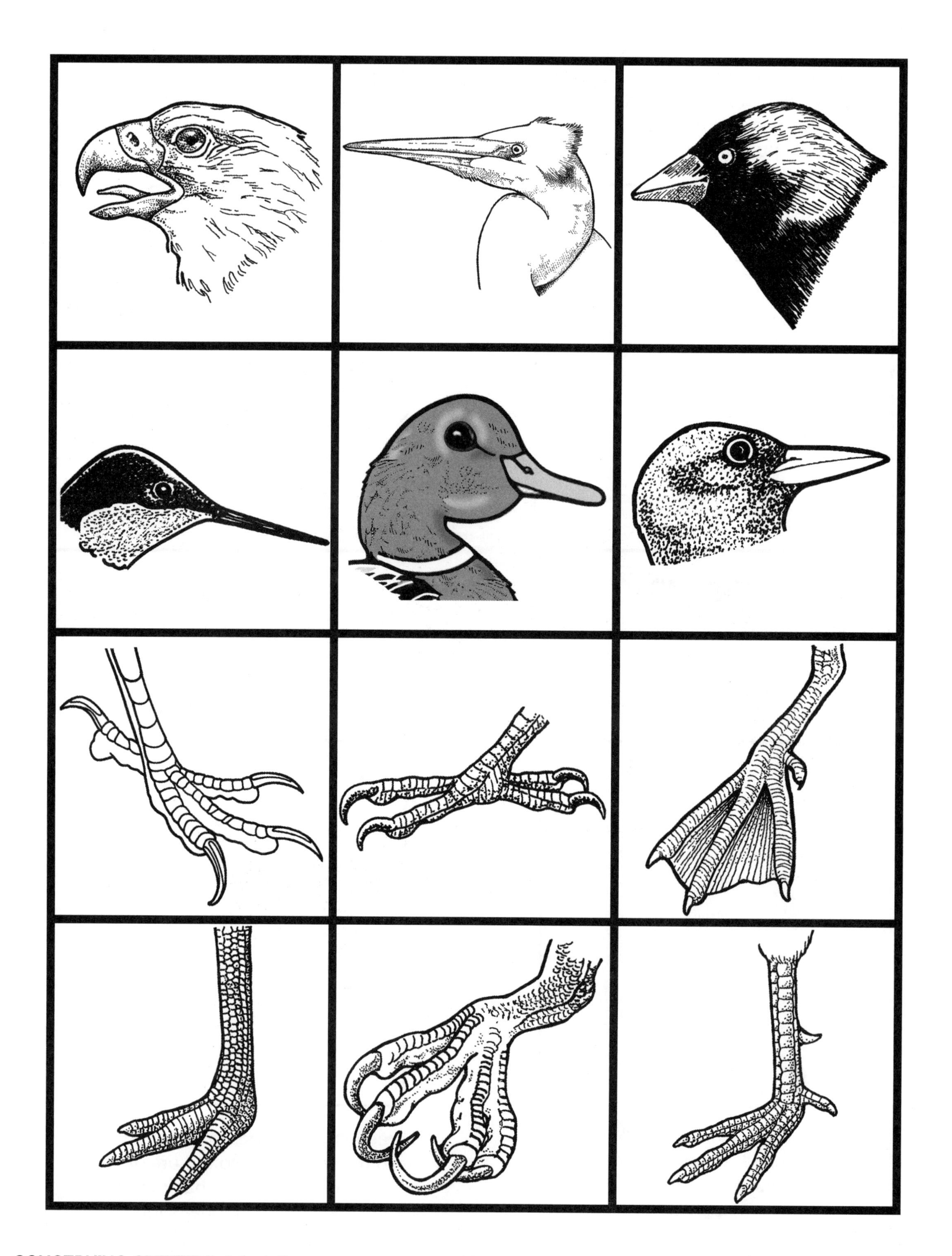

Build a Bird

Build a Bird

Build a Bird

Build a Bird

Build a Bird

Build a Bird

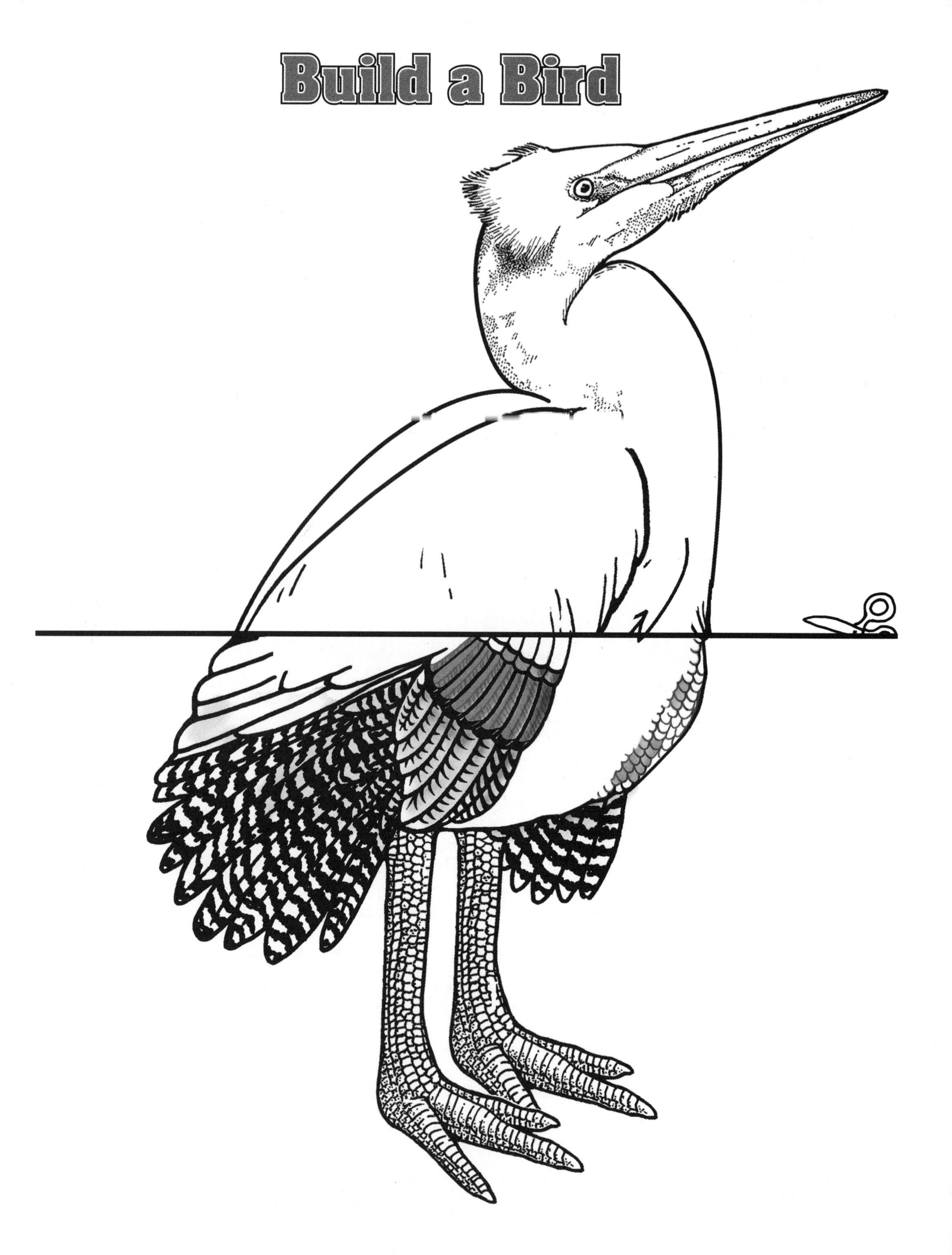

Bird Beaks and Fowl Feet

CONNECTING LEARNING

Connecting Learning

1. What are some of the different kinds of feet that birds can have? ...beaks?
2. What purposes do these different feet and beaks serve?
3. How did you decide which beaks and feet to use for the imaginary birds? Were there ever disagreements? Why or why not?
4. What would you guess that the feet of a sparrow would look like? Why?
5. What kind of beak do you think a teal has? Why?
6. What are you wondering now?

Topic
Adaptations

Key Question
How do the hooves of some prairie animals help them survive?

Learning Goal
Students will observe, compare, and contrast the hooves of prairie animals and relate the structure to their survival.

Guiding Documents
Project 2061 Benchmarks
- *A great variety of kinds of living things can be sorted into groups in many ways using various features to decide which things belong to which group.*
- *Features used for grouping depend on the purpose of the grouping.*
- *For any particular environment, some kinds of plants and animals survive well, some survive less well, and some cannot survive at all.*

NRC Standard
- *Each plant or animal has different structures that serve different functions in growth, survival, and reproduction. For example, humans have distinct body structures for walking, holding, seeing, and talking.*

Science
Life science
  adaptations
    hooves

Integrated Processes
Observing
Comparing and contrasting
Relating

Materials
Student page
Animal photos, included

Background Information
An adaptation is a unique body structure (or behavior) that helps an animal to survive and reproduce in its environment. By looking at the feet and legs of animals, inferences can be made as to their specific purposes. Aquatic animals often have webbed feet that help them swim. Wading birds have long legs. Polar bears have large feet to disperse their weight on thin ice. The toes of a gecko have scales with hooked bristles that help them climb and hold on to any surface. The feet and legs of animals are adapted to help them jump, swim, hunt, dig, climb, defend themselves, and hold things.

This activity is focused on the adaptive structures of the feet of prairie animals, specifically those with hooves. Mammals that have hooves are called ungulates. A hoof is a part of a foot. A hoof is made of keratin, the same material that makes our fingernails. This hard but flexible substance helps protect the animal's feet. There is the plate-like portion and then a softer layer—much like a toenail. However, hooves are the portion of the animal's foot that comes in contact with the ground.

Hooves of prairie animals are found in two basic shapes: cloven and solid. Cloven hooves are divided into two or more parts. Goats, sheep, deer, cattle, pigs, and bison have cloven hooves. These help them to keep their balance on uneven ground where they might need to go for food or to escape their enemies. The hooves of cattle and pigs are divided into four parts, two of which look like back toes. These allow them to walk in muddy areas without sinking down.

Horses, donkeys, and mules have solid hooves. Their hooves are adapted for speed and defense. Animals with solid hooves are relatively large. The flexible hoof of the laden animal will splay out as it walks, thus distributing its body weight over a greater surface. Pioneers used these animals to carry their packs and even themselves across the countryside.

Management
1. You will need a projection device in order to show students the photos found on the CD.

Procedure
1. Ask students what an adaptation is. [A unique body structure (or behavior) that helps an animal to survive and reproduce.] Inform the class that they will be looking at the hooves of animals to determine their functions. Ask the *Key Question* and state the *Learning Goal.*
2. Lead a discussion on why students think that certain animals have hoofs. Tell them that they will learn about two different types of hooves that prairie animals have and how their hooves help the animals survive.

3. Ask students to describe a prairie. [Broad lands that contain flat areas covered with grass. There may be rolling hills that have rocky outcrops. Prairies have few trees.] Invite them to name some animals that live on the prairies.
4. Distribute the student page that describes the two types of hooves. Give students time to read it and study the illustrations.
5. Show the photos of the animals found on the CD. Tell students to focus in on the structure of the hoof. When they have decided what type of hoof structure each animal has, have them record this and its purpose or function on their page.
6. Conclude with a discussion about hoof adaptations.

Connecting Learning

1. What two types of hooves do prairie animals have? [cloven and solid]
2. How do the hooves help the animals to survive? [cloven hooves help animals climb to find food and escape danger, solid hooves are for speed and weight distribution]
3. What type of hooves do you think a mountain goat has? Explain why you think this. [cloven, to climb steep mountains with uneven surfaces]
4. Which type of hoof would you like to have? Why?
5. What other foot adaptations do animals have? [webbed feet for swimming, claws for digging, long toes for grasping and climbing, etc.]
6. What are you wondering now?

Solutions

Bison/buffalo: solid, to run fast and disperse weight
Cow: cloven, to walk/climb on uneven surfaces, to not sink in the mud
Deer: cloven, to walk/climb on uneven surfaces
Donkey: solid, to run fast and disperse weight
Horse: solid, to run fast and disperse weight
Mule: solid, to run fast and disperse weight
Pig: cloven, to walk/climb on uneven surfaces, to not sink in the mud
Sheep: cloven, to walk/climb on uneven surfaces

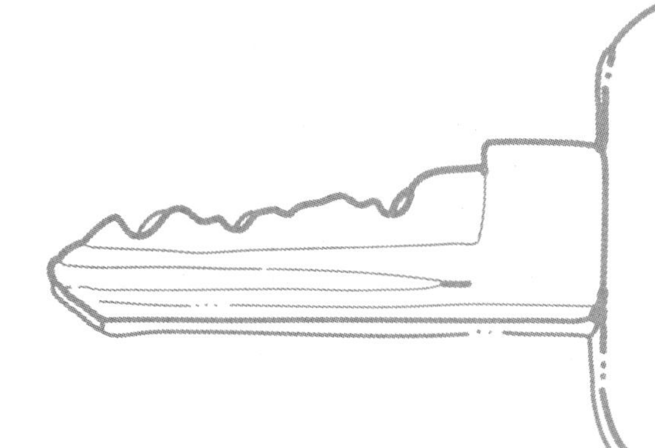

Key Question

How do the hooves of some prairie animals help them survive?

Learning Goal

Students will:

observe, compare, and contrast the hooves of prairie animals and relate the structure to their survival.

Many prairie animals have hooves. Hooves are a part of their feet. Instead of walking on the soft parts of their feet, they walk on their hooves. Hooves are made of a hard material much like our toenails.

There are two types of hooves you will be observing.

SOLID

One is a **solid** hoof. It looks like this. This adaptation helps the animal run fast on the solid prairie ground. It also helps to spread the weight of the animal over a greater surface area. Heavier animals often have solid hooves.

CLOVEN

The other type of hoof is the **cloven** hoof. See how it looks like it is split into parts? A cloven hoof helps animals climb on uneven surfaces like rock outcrops. The animals with cloven hooves may want to graze on food that is on these higher places. They may also want to escape to these areas to get away from predators.

Some animals with cloven hooves have smaller hooves growing on the backs of their legs. These help them when they are walking in muddy areas to not sink too far in the mud.

Record information about the hooves you see on the animals in the photographs.

ANIMAL	HOOF TYPE	PURPOSE

Connecting Learning

1. What two types of hooves do prairie animals have?
2. How do the hooves help the animals to survive?
3. What type of hooves do you think a mountain goat has? Explain why you think this.
4. Which type of hoof would you like to have? Why?
5. What other foot adaptations do animals have?
6. What are you wondering now?

Bear Feet

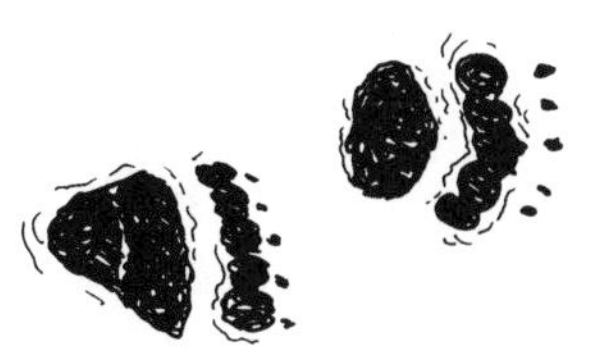

Topic
Adaptations

Key Question
How does a bear's feet help it meet its needs?

Learning Goals
Students will:
- compare and contrast the feet of the polar bear, the brown bear, and the black bear;
- see how the structure fits the function;
- make bear footprint stamps; and
- build dioramas of the bears' habitats.

Guiding Documents
Project 2061 Benchmarks
- *A great variety of kinds of living things can be sorted into groups in many ways using various features to decide which things belong to which group.*
- *Features used for grouping depend on the purpose of the grouping.*
- *For any particular environment, some kinds of plants and animals survive well, some survive less well, and some cannot survive at all.*
- *Individuals of the same kind differ in their characteristics, and sometimes the differences give individuals an advantage in surviving and reproducing.*

NRC Standards
- *An organism's behavior evolves through adaptation to its environment. How a species moves, obtains food, reproduces, and responds to danger are based in the species' evolutionary history.*
- *Biological evolution accounts for the diversity of species developed through gradual processes over many generations. Species acquire many of their unique characteristics through biological adaptation, which involves the selection of naturally occurring variations in populations. Biological adaptations include changes in structures, behaviors, or physiology that enhance survival and reproductive success in a particular environment.*

Science
Life science
adaptations

Integrated Processes
Observing
Comparing and contrasting
Inferring
Applying
Generalizing

Materials
For each student:
1" wooden block or empty film canister with flat-topped lid
moleskin, 1" x 1" piece
fine-line markers, pens, or pencils
materials for diorama
toothpick
sharp scissors

For the class:
ink pads for stamps

Background Information

The National Research Council states in their publication *National Science Education Standards* (National Academy Press, Washington DC. 1996. pp 157-158) that it is important for students to develop the understanding that

> An organism's behavior evolves through adaptation to its environment. How a species moves, obtains food, reproduces, and responds to danger are based in the species' evolutionary history.
>
> Species acquire many of their unique characteristics through biological adaptation, which involves the selection of naturally occurring variations in populations. Biological adaptations include changes in structures, behaviors, or physiology that enhance survival and reproductive success in a particular environment.
>
> Extinction of a species occurs when the environment changes and the adaptive characteristics of a species are insufficient to allow its survival.

An *adaptation* is a characteristic of an organism that helps it to survive and reproduce in a particular environment. Organisms are fitted to their environments when they are adapted to them. Individual organisms do not deliberately adapt; they either have the adaptive ability or they don't. When conditions in the environment change requiring the adapation, those with the adaptive ability survive while those without it will not.

The three North American bears featured in this activity, the polar bear, the black bear, and the brown bear, all belong to the same genus, *Ursus*. Their species, however, differ. One of the distinguishing features of these bears is their feet which, are adapted to permit them to live and survive in their particular environments.

All three bears have plantigrade feet, which means they stand flat-footed on their hind feet. Having plantigrade feet allows the bears to stand upright, as straight as humans. Scientists believe they do this to better smell their surroundings. (Animals like the dog and cat have digitigrade feet; they walk on their toes.) For information about the bears' feet, see the student pages.

The polar bear is known as *Ursus maritimus* which means "bear of the sea." The polar bear virtually spends its life on the ice or in the ocean. It differs from the two other North American bears in that it is essentially carnivorous. The ringed seal and the bearded seal make up the majority of its diet. Along with its distinctive feet, the polar bear has many adaptations that allow it to survive in its frigid environment.

The black bear is known as *Ursus americanus* or "American bear." It is the smallest in size of the North American bears. However, it is the largest in numbers. Black bears have adapted to various habitats—forests, meadows, and wetlands—but they never stray too far from the forest, where much of their food and protection can be found. Black bears prefer nuts and seeds, but will eat just about anything edible—berries and other fruits, greens, insects, small rodents, and human garbage, for which they are often considered pests. Black bears are the least aggressive of the North American bears.

The brown bear is classified as *Ursus arctos*. *Ursus* is Latin for "bear" and *arctos* is Greek for "bear." The brown bear, or "bear bear," is certainly what we consider the epitome of a bear. It is immense in size with massive muscles, long claws, a short temper, and tremendous appetite. The North American brown bears inhabit mountains and northern wilderness. They are considered open-space animals. The grizzly bear (*Ursus arctos horribilis*), which lives west of the Mississippi River in the contiguous US and throughout Canada and Alaska, and the kodiak bear (*Ursus arctos middendorffi*), that lives on Kodiak Island in Alaska, are two types of brown bears. They are omnivores, which seems contradictory to the *Carnivora* order to which they belong. Brown bears are opportunists, eating what is seasonally abundant. Their major sustenance is from plant matter. They seldom hunt, but eat carrion, small rodents, and fish when available.

(*Our thanks to Burleigh Lockwood, Educational Biologist and Presenter at the Chaffee Zoo in Fresno, CA, for her consultation on this activity.*)

Management

1. Moleskin can be found in the foot care section of drug stores. Dr. Scholl's in one brand name that is available.
2. To make the stamps, the moleskin can be applied to the wooden cube or to the lid of the empty film canister. If students are making right and left paw prints, one can be put on the lid of the canister with the other on the bottom end of the canister. If using the blocks, adhere the stamps to two different faces of the cube.

3. Students can use fine-lined markers, pencils, or pens to add details to the bears' footprints. Toothpicks can be used to help separate the paper covering from the moleskin.
4. Have students work in groups of three with each student making a footprint stamp and a diorama depicting the habitat of one of the three bears.
5. Shoe boxes make excellent structures for dioramas. Have construction paper available. Encourage students to bring in natural items (twigs, grasses, pebbles, etc.) to add to the habitats.
6. Salt-flour clay can be used to form the ice and snow or mountains for the various habitats. The clay can also be used if students want to make models of the bears. Salt-flour clay is made by mixing two parts flour with one part salt and adding enough water to make a stiff dough. It can be air-dried or baked at low temperatures and painted.
7. Materials should be made available so that students can do further research on the three bears.

Procedure

1. Distribute the information pages for students to read.
2. Ask the *Key Question*.
3. Have students illustrate the feet of the three different bears and describe how the structure fits the function.

4. Have students make stamps of bears' feet. Once they have inked their stamp and stamped tracks onto paper, urge them to fill in necessary details with pencils, markers, or pens.
5. Encourage them to make dioramas of the bears' habitats and use the stamps to make bear tracks. It should be stressed that students make as complete a model of the three habitats as supplies will allow to illustrate the function of the bears' feet. For example, the polar bear's habitat should show the ice and waters of the Arctic with perhaps a partially consumed seal's carcass. The black bear's habitat should be one of forests with berries, nuts, and seeds. Strips of peeled bark can be used to depict the bark that black bears strip in order to reach the tasty cambium layer of the tree. The brown bear's habitat should be open meadows perhaps using corrugated cardboard to illustrate the ground the bear has "plowed" as it has looked for roots.

Connecting Learning
1. How are the three bears' feet alike? How are they different?
2. How would you distinguish a polar bear's footprint from that of a brown bear's?
3. Which footprints would be the most difficult to distinguish? [the brown bear's and the black bear's] Why?
4. What would be some clues that would help you to know? [the brown bear's front claws are very long and the toe marks appeared to be joined]
5. Describe the footprint you might find that leads to the base of a tree but nowhere else. [pad with toes that are arced with short claws] To which bear do the prints probably belong? [black bear] What caution do you need to take? [Look up in the tree, it may have climbed up there.]
6. What other adaptations of the three types of bears would you like to investigate?
7. What are you wondering now?

Curriculum Correlation

Gilly, Shelley. *Alaska's Three Bears.* Paws IV. Publishing Company. Homer, AK. 1990.

Hodge, Deborah. Bears: *Polar Bears, Black Bears and Grizzly Bears.* Kids Can Press. Toronto, ON. 1996.

Rosing, Norbert. *Face to Face With Polar Bears.* National Geographic. Washington, DC. 2009.

Sartore, Joe. *Face to Face With Grizzlies.* National Geographic. Washington, DC. 2009.

Swinburne, Stephen R. *Black Bear: North America's Bear.* Boyds Mills Press, Inc. Honesdale, PA. 2003.

Science
1. Have students investigate the friction of different materials on ice.
2. Have students compare and contrast various types of shoe soles. Urge them to determine how the design fits the function. Continually refer to how the bears' feet fit the various functions necessary for its survival.

Social Studies
1. Have students locate and distinguish the areas of the various bears.

Math
1. Have students join sheets of chart paper and draw the bears according to researched average length and height.
2. Have students use grid paper to make scaled drawings of the footprints.

Bear Feet

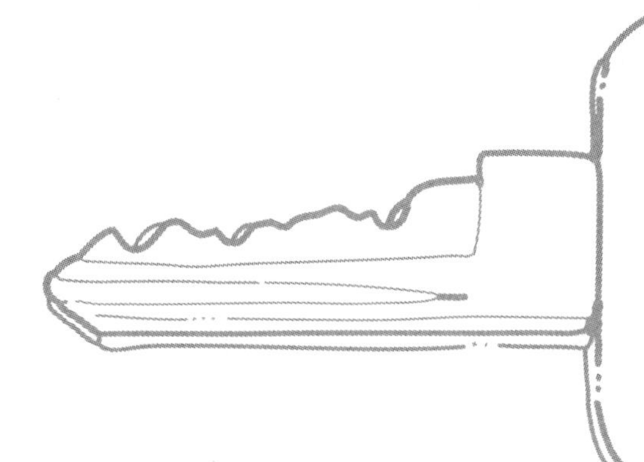

Key Question

How does a bear's feet help it meet its needs?

Learning Goals

Students will:

- compare and contrast the feet of the polar bear, the brown bear, and the black bear;
- see how the structure fits the function;
- make bear footprint stamps; and
- build dioramas of the bears' habitats.

Bear Feet Polar Bear

The polar bear is known as *Ursus maritimus* which means "bear of the sea." It is found in the far northern parts of the world, an environment of ice, snow, and frigid waters. The polar bear is very comfortable in this environment.

Along with many other adaptations that enable it to survive, the polar bear's huge feet help to distribute its weight over a large area, allowing it to walk on thin ice. The soles of its feet have thick leather pads with fur between the toes. The friction created by the fur and pads helps prevent the polar bear from sliding as it walks on the ice and snow. The front feet are slightly webbed to aid it as it swims. The back feet and powerful hindquarters are used for steering in the water. The front feet are equipped with sharp, curved claws that can grasp and pull a seal out through the hole in the ice the seal uses for breathing. The seal then becomes the polar bear's meal.

Bear Feet Black Bear

The black bear is known as *Ursus americanus* or "American bear." It is the smallest in size of the North American bears. However, it is the largest in numbers. Black bears have adapted to various habitats, forests, meadows, and wetlands, but they never stray too far from the forest where much of their food and protection can be found.

The black bear is a natural climber with feet that have leathery soles that provide traction on the bark of trees. The curved, blunt, and dark-colored claws found on all four of its feet enable it to quickly climb trees and dig for food. Its claws are arranged in an arc formation much like the fingers on our hands. It is thought that this arrangement helps it to grasp the bark of the trees as it climbs. It also uses its claws as tools to strip the bark from certain trees so it can eat the juicy outside layers.

Bear Feet Brown Bear

The most familiar North American brown bears (*Ursus arctos* which means "bear bear") are grizzlies (Ursus *arctos horribilis*) and kodiaks (*Ursus arctos middendorffi*). Brown bears are open-space animals which find their needs met in meadows, river valleys, and mountains. Although they belong to the *Carnivora* order, they are omnivores. Their diet consists mostly of plants: roots, fruit, wild berries, nuts, flowers, etc. However, they are excellent fishers and have been known to catch and partially consume some 20 salmon per hour during prime salmon fishing season.

Brown bears have heavy pads on the soles of their feet. Their toes are arranged straight across the tops of their feet much like the toes on our feet. Their front claws may be 15 cm in length. These long claws can delicately pick up tiny berries, but combined with the huge shoulder muscles, they can also turn a meadow into what looks like a plowed field when digging for roots. The length of the adult brown bears' claws makes it difficult for them to climb trees. The footprint of a brown bear can be distinguished from that of a black bear by looking for evidence of long front claws and joined toe marks.

Bear Feet

Illustrate a foot of each bear. Describe how the features of the feet help the animals survive.

Polar Bear

Black Bear

Brown Bear

Bear Feet

Connecting Learning

1. How are the three bears' feet alike? How are they different?
2. How would you distinguish a polar bear's footprint from that of a brown bear's?
3. Which footprints would be the most difficult to distinguish? Why?
4. What would be some clues that would help you to know?
5. Describe the footprint you might find that leads to the base of a tree but nowhere else. To which bear do the prints probably belong? What caution do you need to take?
6. What other adaptations of the three types of bears would you like to investigate?
7. What are you wondering now?

Topic
Adaptations

Key Question
How do bats keep warm in a cave?

Learning Goal
Students will explore how some bats have adapted their behavior to meet their needs in cool cave environments.

Guiding Documents
Project 2061 Benchmark
- *For any particular environment, some kinds of plants and animals survive well, some survive less well, and some cannot survive at all.*

NRC Standard
- *An organism's patterns of behavior are related to the nature of that organism's environment, including the kinds and numbers of other organisms present, the availability of food and resources, and the physical characteristics of the environment.*

Math
Measurement
  temperature

Science
Life science
  adaptations
  organisms
  habitats

Integrated Processes
Observing
Comparing and contrasting
Collecting and recording data
Applying

Materials
Per class or group (see Management):
  thermometer
  strand of 50 or more miniature lights
  shoebox or other similar-sized box with a lid
  bag of cotton balls (at least 300)
  white glue
  one sheet of 8.5 x 11-inch cardboard or similar material

Background Information
Many species of bats roost in caves all or part of the year. Since the ambient temperature in these caves is usually significantly less than the bats' normal (non-hibernating) body temperatures, these bats will often "bunch" together to keep warm. This behavioral adaptation to the cool cave environment is especially important for the newborn bats of some species, like Mexican free-tails. These babies, which are called pups, are born hairless and must keep their body temperature around 100 degrees Fahrenheit in order to live. In his book, *America's Neighborhood Bats* (1988), bat expert Merlin Tuttle notes that after free-tailed pups are born they will "roost in nearly solid masses at densities sometimes in excess of 500 per square foot, covering hundreds of feet of cave walls." The warmth produced by this bunching helps keep the pups alive in the cool ambient temperatures of their cave environment.

Management
1. This activity is written so that it can be done as a whole class or in groups.
2. If you want to do the activity in groups, each group needs its own strand of miniature lights (like the ones used on Christmas trees) and a large bag (more than 300) of cotton balls. Note: Do not use LED lights, which put out very little heat. Use incandescent bulbs.
3. **Caution:** In *Part One* when students huddle together, they must **not** push or shove.
4. **Caution**: The strands of miniature lights used in *Part Two* may become hot if left in the box too long. Make sure they are not plugged in for more than five minutes at a time.

5. When the strand of miniature bulbs is plugged in, all the bulbs will be lit. Only the requisite number (1, 10, 50) of bulbs are placed in the box for each temperature measurement. (**The lights inside the box need to be placed so they don't touch the thermometer.**) The rest of the lights on the string should be placed so that they don't touch the box, thus adding to its internal temperature. The easiest way to do this is to have the strand exit the box at a corner. If needed, a small hole can be cut in one corner for the cord.
6. Thermometers (item number 1976) are available from AIMS.

Procedure

Part One

1. Ask students if they've ever been inside a cave. Ask those who have to describe the cave's temperature.
2. Ask the *Key Question:* How do bats keep warm in a cave?
3. Distribute the first student page, and have students share their theories on how bats stay warm in the coolness of their cave environment and then record these theories in the space provided.
4. Tell students they are going to experience first-hand how bats stay warm by huddling together as closely as they can.
5. Divide your class into two groups—boys and girls. Have them form two "bunches" in separate areas of the classroom. Ask students to huddle as closely as they can to each other without pushing or shoving. Remind them to keep their hands down and to themselves as they do this. After a few minutes, have students return to their seats.
6. Have students discuss what happened to the temperature as they huddled together and then complete the first page.

Part Two

1. Distribute the second student page and the thermometers, boxes, and lights. Demonstrate how the requisite number of lights can be placed in the box in such a way that the inside lights do not touch the thermometer and the outside lights don't touch the outside of the box. (It's okay if the inside lights touch the interior of the box.)
2. Have students follow the instructions on the page as they gather temperature data with the miniature lights.
3. After all the data are collected, discuss the results.

Part Three

1. Distribute the final student page, the piece of cardboard, the glue, and the cotton balls to each group.
2. Have students count out 300 cotton balls.
3. Using a single drop of glue for each ball, have students attach all 300 to the sheet of cardboard.

Connecting Learning

Part One

1. What is the temperature like inside a cave? [Ambient temperatures inside most caves are cool, usually in the 50s and 60s Fahrenheit.]
2. How do bats keep warm inside a cave? [Bats are warm-blooded and expend internal energy to stay warm. Many species of bats will huddle closely together to conserve energy.]
3. What happened to the temperature as you huddled closely together? [The temperature went up and it felt warm or hot, depending on where in the huddle you were.]
4. How does this huddling activity help answer the question of how bats stay warm inside a cave? [Bats bunched closely together keep each other warm.]
5. What do you think would happen if the bats did not huddle together?
6. How do you think the mother bat finds her pup in the bat nursery? [She uses her sense of hearing to detect its sound. She also uses her sense of smell. Scientists also believe she "remembers" its location.]

Part Two

1. What happened to the temperature when the various numbers of lights were placed in the box? [The more lights that were placed in the box, the higher the temperature went.]
2. How does this model what happens inside a bat cave? [The ambient temperature of the cave rises as the many bats roosting there heat it. The light bulbs are like the warm-blooded bats in the cave—both give off heat.]

Part Three

1. What did you learn about bats as you glued the cotton balls to the piece of cardboard? [The bat "pups" represented by the cotton balls are bunched very close together.]
2. What are some of the things you learned after doing the three parts of this activity? [Some bats huddle together for warmth. The temperatures in caves are cool. Bat pups are hairless. Bunching together for warmth in a cool cave is adaptive behavior. Etc.]
3. What other animals have you seen huddle together to keep warm? [penguins, puppies, kittens, cows in a herd, etc.]
4. What are you wondering now?

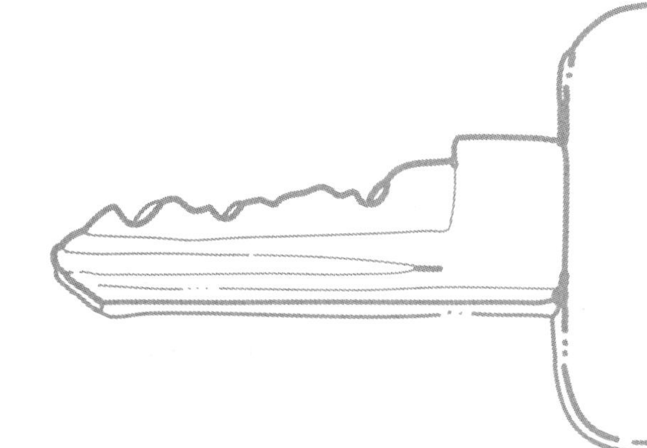

Key Question

How do bats keep warm in a cave?

Learning Goal

Students will:

explore how some bats have adapted their behavior to meet their needs in cool cave environments.

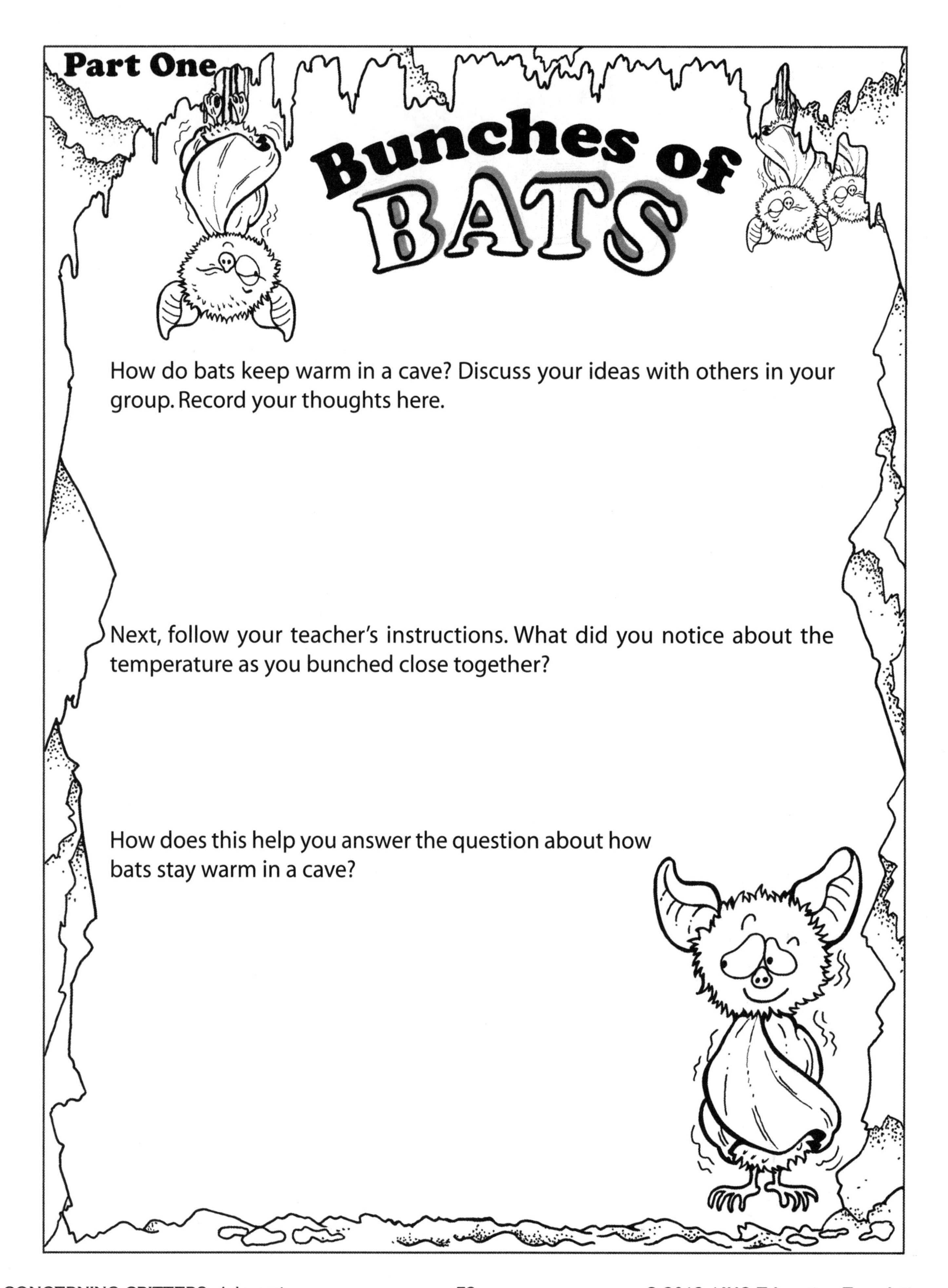

How do bats keep warm in a cave? Discuss your ideas with others in your group. Record your thoughts here.

Next, follow your teacher's instructions. What did you notice about the temperature as you bunched close together?

How does this help you answer the question about how bats stay warm in a cave?

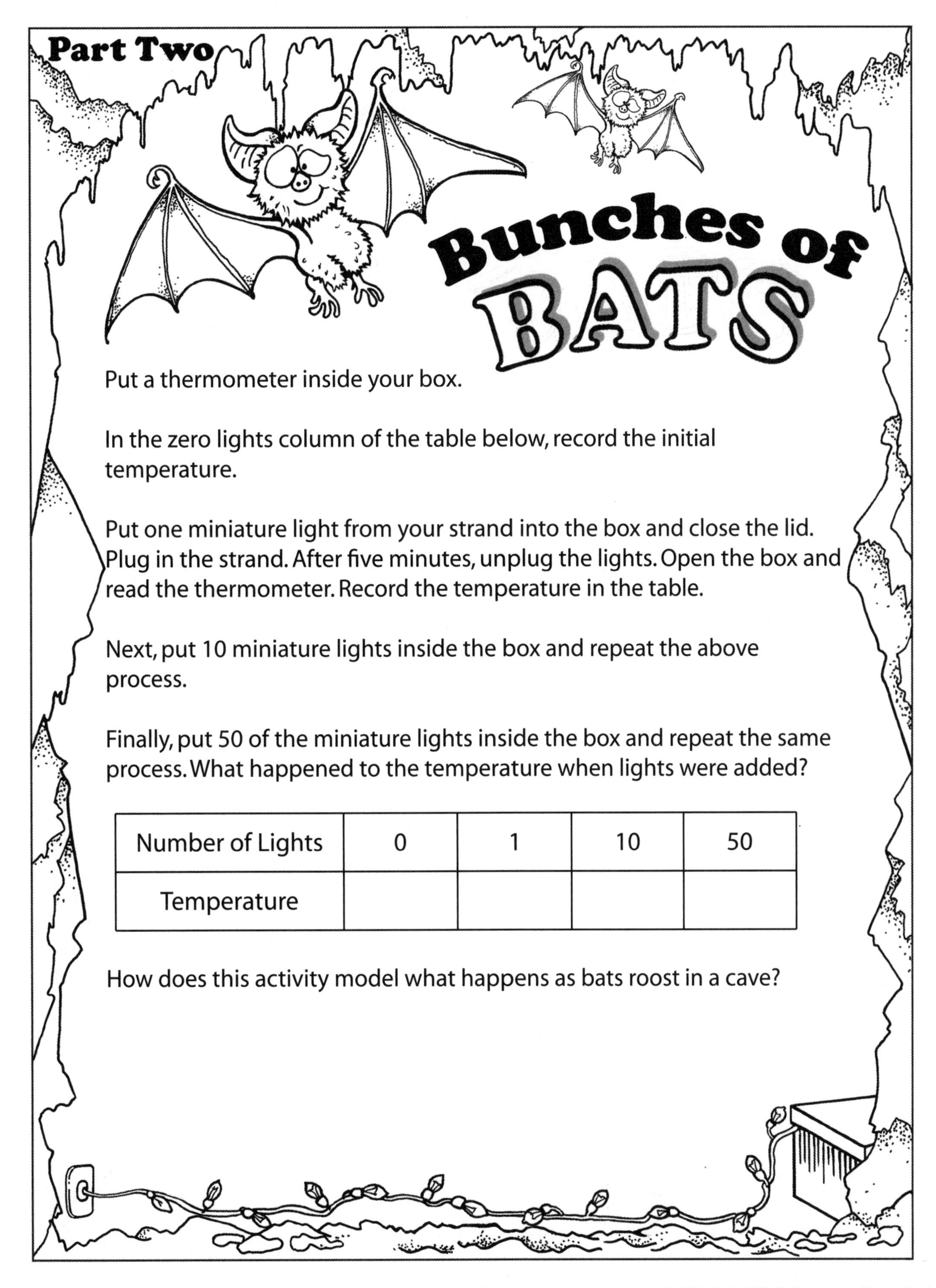

Bunches of BATS

Put a thermometer inside your box.

In the zero lights column of the table below, record the initial temperature.

Put one miniature light from your strand into the box and close the lid. Plug in the strand. After five minutes, unplug the lights. Open the box and read the thermometer. Record the temperature in the table.

Next, put 10 miniature lights inside the box and repeat the above process.

Finally, put 50 of the miniature lights inside the box and repeat the same process. What happened to the temperature when lights were added?

Number of Lights	0	1	10	50
Temperature				

How does this activity model what happens as bats roost in a cave?

Part Three

Bunches of BATS

Baby bats are called pups. The pups of some species, like Mexican free-tailed bats, are born hairless and must maintain a body temperature of about 100 degrees Fahrenheit in order to survive. To do this, they bunch together so closely that more than 500 of them cover a single square foot of cave wall. This means that about 300 of these pups would fit on this sheet of paper!

Count out 300 cotton balls. Each ball represents one Mexican free-tailed bat pup. Using a single drop of glue for each ball, work with your group to attach all 300 "pups" to your sheet of cardboard. This models how closely together the pups are bunched as they roost on cave walls.

In the space below, tell what you've learned in this activity.

Connecting Learning

Part One

1. What is the temperature like inside a cave?

2. How do bats keep warm inside a cave?

3. What happened to the temperature as you huddled closely together?

4. How does this huddling activity help answer the question of how bats stay warm inside a cave?

5. What do you think would happen if the bats did not huddle together?

6. How do you think the mother bat finds her pup in the bat nursery?

CONNECTING LEARNING

Connecting Learning

Part Two

1. What happened to the temperature when the various numbers of lights were placed in the box?

2. How does this model what happens inside a bat cave?

Part Three

1. What did you learn about bats as you glued the cotton balls to the piece of cardboard?

2. What are some of the things you learned after doing the three parts of this activity?

3. What other animals have you seen huddle together to keep warm?

4. What are you wondering now?

Nocturnal Hunter

Topics
Adaptations/Food chains

Key Questions
1. What adaptations does an owl have that help it to catch its prey?
2. What can you learn about food chains by studying an owl pellet?

Learning Goals
Students will:
- learn about owl adaptations;
- dissect an owl pellet to discover the diet of an owl;
- draw, organize, and identify the bones recovered; and
- construct a food chain.

Guiding Documents
Project 2061 Benchmarks
- *Keep records of their investigations and observations and not change the records later.*
- *Almost all kinds of animals' food can be traced back to plants.*

NRC Standards
- *Use appropriate tools and techniques to gather, analyze, and interpret data.*
- *Populations of organisms can be categorized by the function they serve in an ecosystem. Plants and some micro-organisms are produces—they make their own food. All animals, including humans, are consumers, which obtain food by eating other organisms. Decomposers, primarily bacteria and fungi, are consumers that use waste materials and dead organisms for food. Food webs identify the relationships among producers, consumers, and decomposers in an ecosystem.*
- *For ecosystems, the major source of energy is sunlight. Energy entering ecosystems as sunlight is transferred by producers into chemical energy through photosynthesis. That energy then passes from organism to organism in food webs.*

Math
Measurement
  length
Whole number operations

Science
Life science
  adaptations
  food chains

Integrated Processes
Observing
Comparing and contrasting
Collecting and recording data
Organizing data
Classifying
Analyzing

Materials
For each group:
  one owl pellet (see *Management 1)*
  forceps
  toothpicks or wooden probes
  metric ruler
  egg carton to separate bones
  bone identification cards
  dish of warm water (see *Management 3)*

For each student:
  hand lens
  student pages
  Owls rubber band book
  #19 rubber band

Background Information
Owls are carnivorous predators at the top level of their food chain. They hunt at night and have many adaptations that allow them to be successful in catching fast-moving prey in the dark. Their large eyes allow a lot of light to enter, giving them excellent night vision. The owl's ears are also crucial to hunting and can be relied on in the darkness. Unlike humans, owls do not have external ears. Instead, the feather arrangement of the facial disk helps gather sound waves and directs them to the eardrum within the skull. Owls' left and right ears are at different levels. This means they can catch sounds at slightly different times, allowing them to better distinguish the location of prey. Owls' feathers are also adapted to help them catch prey. Most birds have stiff feathers that make noise when they fly. The owl's feathers are fringed. This reduces air disturbance and cuts down on noise.

Because of this, an owl's prey does not often hear it coming. Once prey is captured, the owl's legs are also protected from bites by the feathers that go all the way down the legs.

Food chains show links between organisms and can be traced back to energy from the sun. Owls produce and regurgitate a pellet that contains the indigestible remains of the animals that they consume. These pellets can be used to determine what the owl ate. An owl's diet consists of rodents and small birds. Evidence of the rodent's/small bird's diet of grasses or seeds may also be found in the owl's pellet. The focus of this experience is for students to learn about owls' adaptations and to gather data to help them construct an example of an actual food chain involving an owl.

Management

1. Owl pellets can be collected in barns or other roosting areas. To sterilize the pellets, put them in a bag with two or three moth balls for three days or microwave them. They can also be ordered through science supply companies (see *Resources).*
2. Students should work together in pairs or small groups. This promotes communication and cooperative learning as well as requiring fewer pellets.
3. It may be necessary to soak sections of pellets in a dish of warm water to separate dried fur and feathers.
4. If any students are sensitive to dander, a simple face mask can be fashioned from a coffee filter and string. The coffee filter can cover the student's nose and mouth. Taping strings to both sides allows the mask to be tied.
5. Hand lenses (item number 1977), forceps (item number 3065), and metric rulers (item number 1909) are available from AIMS.

Procedure

1. Ask the *Key Question* and state the *Learning Goals.*
2. Distribute the *Owls* rubber band book to students. Read and discuss the information it contains as a class.
3. Tell the students to form groups based on the number of pellets available.
4. Distribute the materials and the student pages.
5. Have students draw, observe, and measure their pellets, recording the requested information on the student page.
6. Using the tweezers and toothpicks, have students take apart the pellets and separate the bones of the animals from fur and feathers. If necessary, instruct students to soak the pellets in the water to remove the fur and feathers from the bones.
7. Once all bones are removed, discard the fur and feathers and have students carefully clean off the bones and separate them according to type. Tell the students to use the bone identification cards.
8. Have students sketch each type of bone and record the number found in the table provided.
9. Using the cards provided, identify the prey consumed based on the skulls and bones discovered.
10. Direct each group to use the bones found to reconstruct a skeleton. Have them sketch in any missing bones using the reference cards and generic rodent skeleton.
11. Ask each student group to construct a food chain based on the evidence they found in the pellet. The chain starts with the sun and ends with the owl. They may need to make some inferences if they found no plant materials in their pellet.

Connecting Learning

1. How were the owl pellets formed?
2. What adaptations help the owl catch its prey?
3. What is the average number of prey found within a pellet?
4. Studies have found that a typical owl regurgitates two pellets per day. Based on this, how many animals do you think an owl eats each day?
5. How many animals would an owl eat in one year?
6. If an owl couple had four young owlets that must be fed for seven weeks until they mature, how much would the rodent population be reduced in this time?
7. What evidence did you find in the pellets that would tell you about the diet of the prey? (Students generally can find plants and seeds with the pellets.)
8. What "chain" can you connect between the sun and an owl, using the evidence you found in the pellet? [sunlight to plants to rodent to owl]
9. What are you wondering now?

Extensions

1. Create a food web in which the barn owl is at the highest tropic level. The food web could be illustrated with photographs connected by string, a diagram drawn on paper, or pictures cut from magazines and arranged in a pyramid.
2. Make a chart of all the bones found by identifying and grouping them. Glue them to a piece of tagboard and give each the appropriate label.
3. Have students do additional research on owls and their behavior.

Curriculum Correlation

Language Arts
Have students write a nonfiction narrative about a typical day in the life of an owl.

Literature
Avi. *Poppy.* Harper Trophy. Toronto. 1997.

Art
Using photographs as references, have students draw owls, paying close attention to detail so they are anatomically correct. The drawings could be incorporated with the narratives and made into a book.

Resources

Owl pellets can be purchased from the following supply companies:

Carolina Biological Supply Company
http://www.carolina.com
1-800-334-5551

Connecticut Valley Biological
http://www.ctvalleybio.com
1-800-628-7748

Flinn Scientific, Inc.
http://www.flinnsci.com
1-800-452-1261

Ward's Natural Science
http://wardsci.com
1-800-962-2660

Nocturnal Hunter

Key Question

What can you learn about food chains by studying an owl pellet?

Learning Goals

Students will:

- dissect an owl pellet to discover the diet of an owl;
- draw, organize, and identify the bones recovered; and
- construct a food chain.

1

Owls are nocturnal. This means they are active at night. They have specially designed eyes, ears, claws, and wings that help them be experts at catching prey.

2

After three weeks, the owlets can swallow prey whole. It takes nine weeks after hatching for the young owls to take their first flights. They first practice with short distances, but when they can fly confidently, their parents stop feeding them. The population of prey in the area has been depleted significantly in those nine weeks, so the parents are anxious for the young to establish territories of their own.

11

Barn owls can live up to 20 years, but few last this long in the wild because of harsh weather or shortage of food.

12

Owls' eyes are bigger than those of most other animals. These large eyes allow a lot of the available light to enter. This gives them amazingly good night vision. Owls also have a large number of rods (light sensing structures) in their eyes. In fact, there is no room in their eyes for muscles. This means that an owl's eyes are fixed in its sockets. To compensate for this, owls can rotate their heads almost 360°. They can easily look directly behind themselves or over either shoulder.

3

4

In addition to lacking muscles, owls' eyes lack cones (color sensing structures). This means they see in black and white. Owls do have three-dimensional vision, which helps them locate prey.

9

Barn owls chiefly feed on voles, rats, shrews, and birds. Owls have their own territories, or hunting areas. Sometimes they will share the same area with other owls if they are different species and do not have the same prey.

After mating, the female owl will lay a clutch of eggs at two-day intervals. This way, not all the chicks will hatch at once, easing the feeding burden. The number of eggs an owl lays depends on the supply of food. It takes 31 days for an owlet to hatch. The babies must be fed regularly, so the hen owl stays near the nest, tearing up the prey the male brings in and feeding it to the young.

10

The owl's ears are also crucial to hunting. They can be relied on in total darkness. Unlike humans, owls do not have external ears. Instead, the feather arrangement of the facial disk helps gather sound waves and directs them to the eardrum within the skull. Owls' left and right ears are at different levels. This means they can catch sounds at slightly different times. This gives the owl one more way to easily locate its prey.

An owl's claws are called talons. They are sharp and curved. There are four toes on each foot. At the moment of attack, an owl fully spreads all eight toes to better grab its prey. Most owls have feathers all the way down their legs to protect them from bites once they have captured their prey.

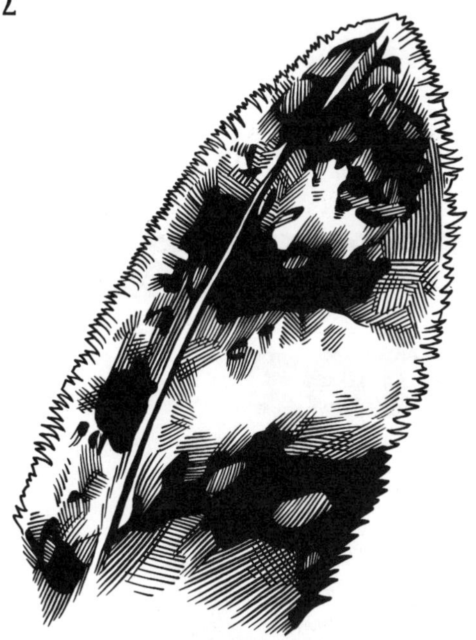

Most birds have stiff feathers that make noise when they fly. The owl's feathers are fringed. This reduces air disturbance and cuts down on noise. Because of this, an owl's prey often does not hear it coming.

After capturing its prey, an owl flies back to its roost to eat. It will usually swallow its prey whole. If the prey is too large, the owl can tear it into smaller pieces with its sharp, curved beak. The owl's digestive juices cannot break down the fur, feathers, or bone that it swallows. Those indigestible materials are clumped together in the stomach and then regurgitated through the mouth. These clumps are called pellets. They can be very useful in determining an owl's diet and the distribution of rodents in a specific area.

Nocturnal Hunter

Examine and record

Length of Pellet	
Width (at widest point)	
Coloration	
Visible Contents	

Make a scale drawing of your pellet.

							1 cm

Now, carefully pick off all of the fur and feathers and discover the contents of the pellet. Separate and clean all bones, even the tiniest ones. Sort them by type: skulls, femurs, vertebrae, etc.

Nocturnal Hunter

RECORD YOUR FINDINGS

SKETCH OF BONE	NUMBER FOUND

Based on the number of skulls found, how many animals were consumed?

____

What is the class average of number of skulls found?

____

An owl often produces two pellets a day. Using your average consumption number, determine how many animals are consumed:

per day: ____________________

per week: ____________________

per month: ____________________

per year: ____________________

Nocturnal Hunter

USING SKULLS FOUND, IDENTIFY PREY CONSUMED

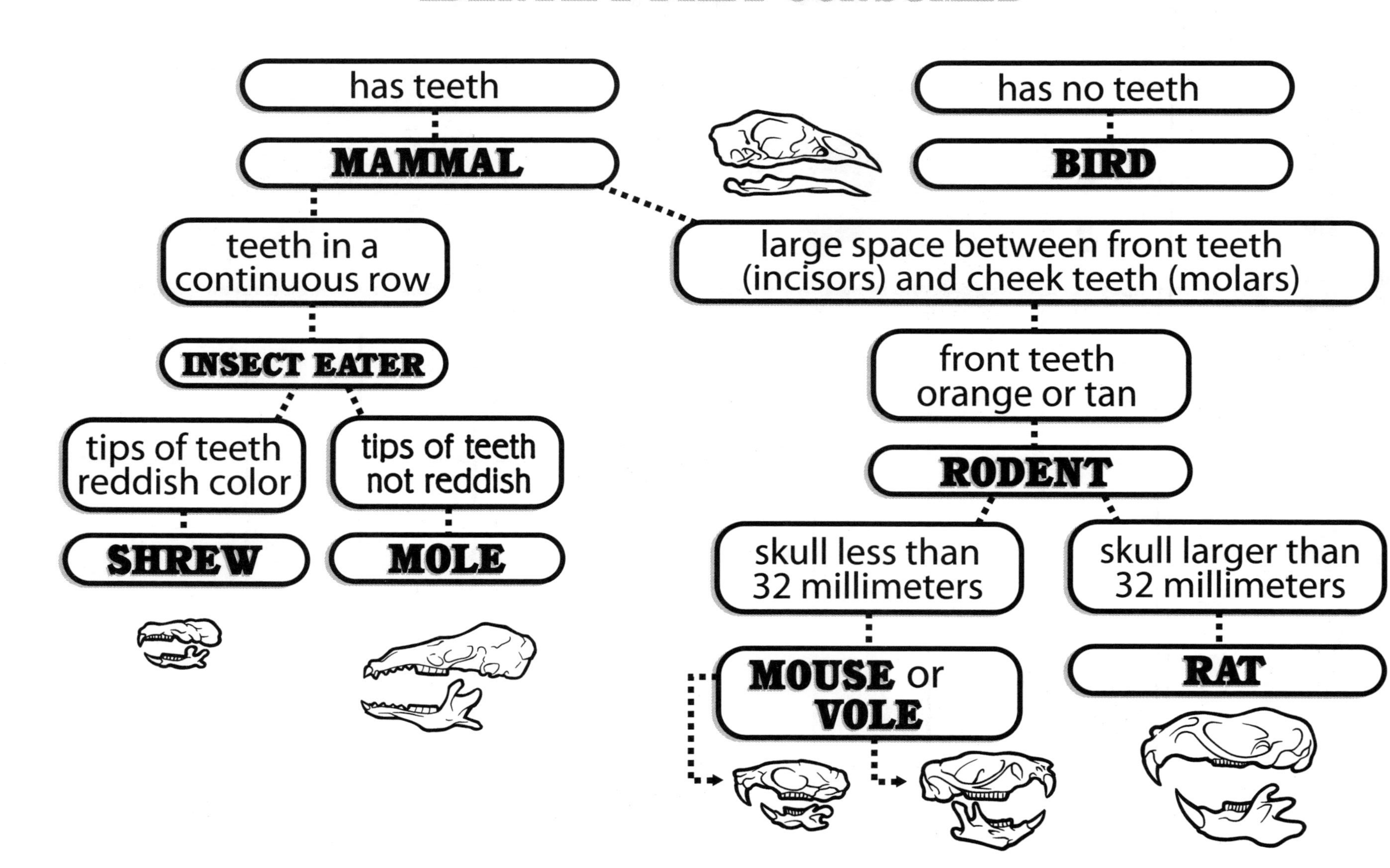

Back
Bones
Rib
Upper
Leg
Shoulder
Blade
Skull
and Jaw
Lower
Leg
Hip
BIRD
Foot

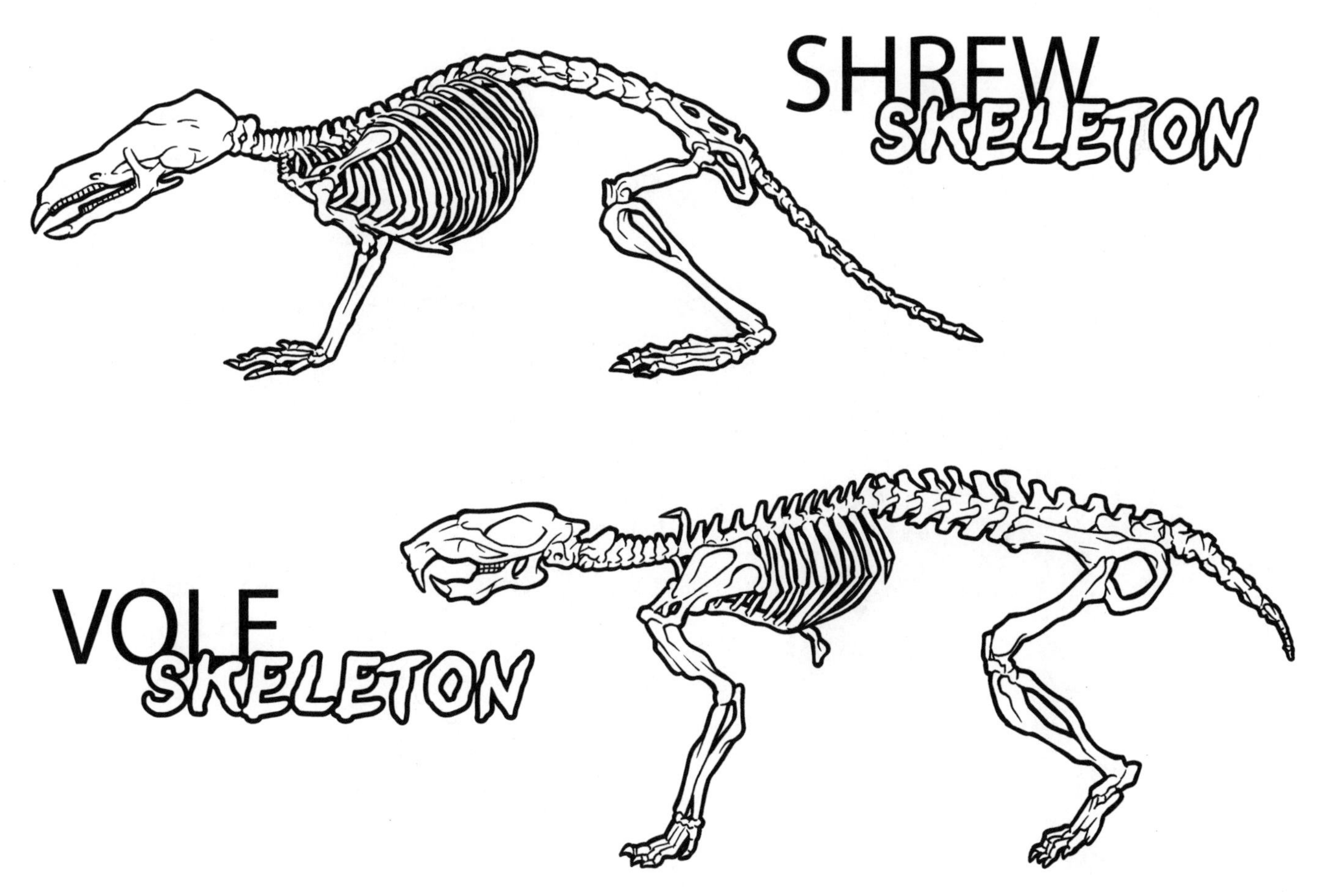
SHREW
SKELETON
VOLE
SKELETON

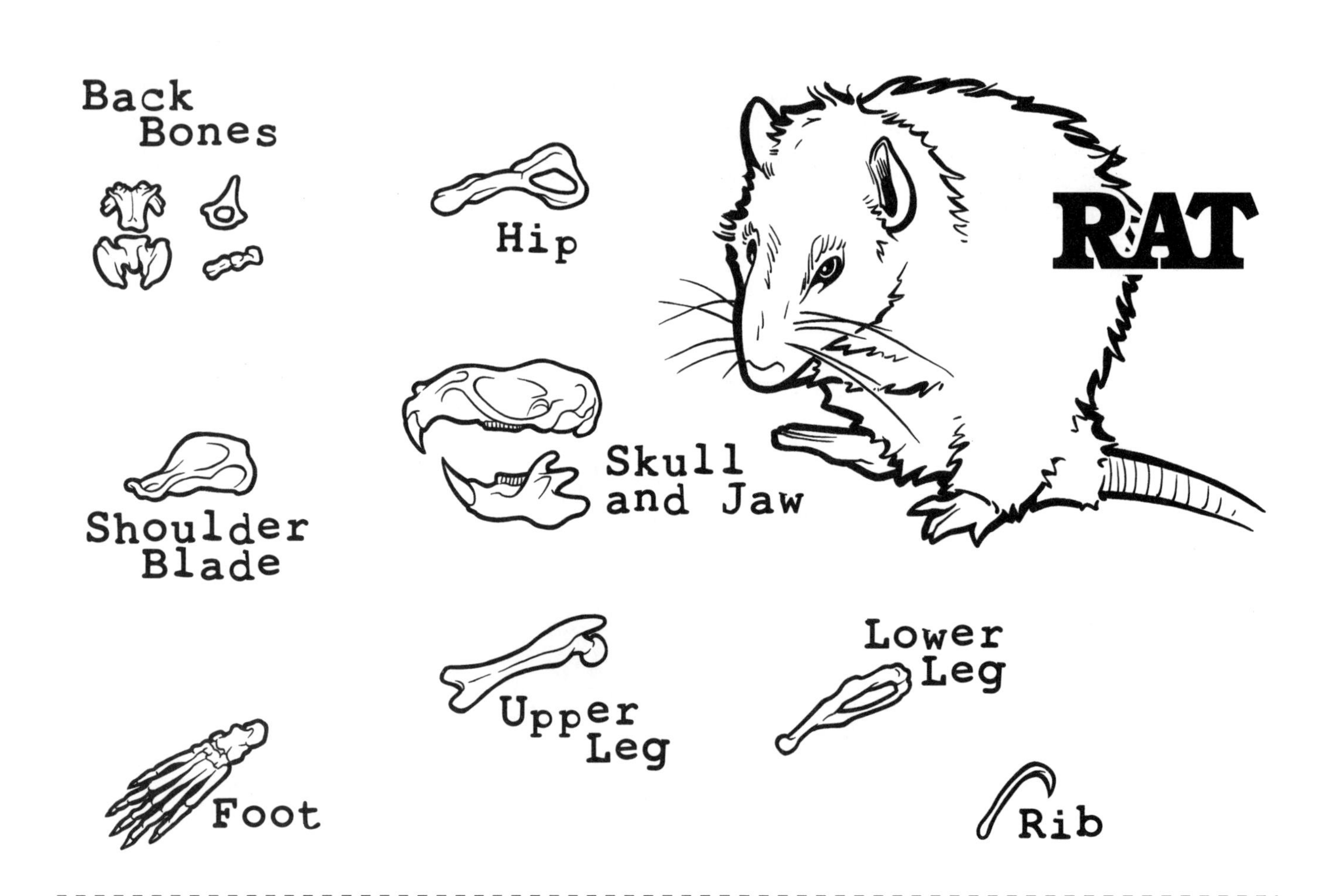
Back
Bones
Hip
RAT
Skull
and Jaw
Shoulder
Blade
Upper
Leg
Lower
Leg
Foot
Rib

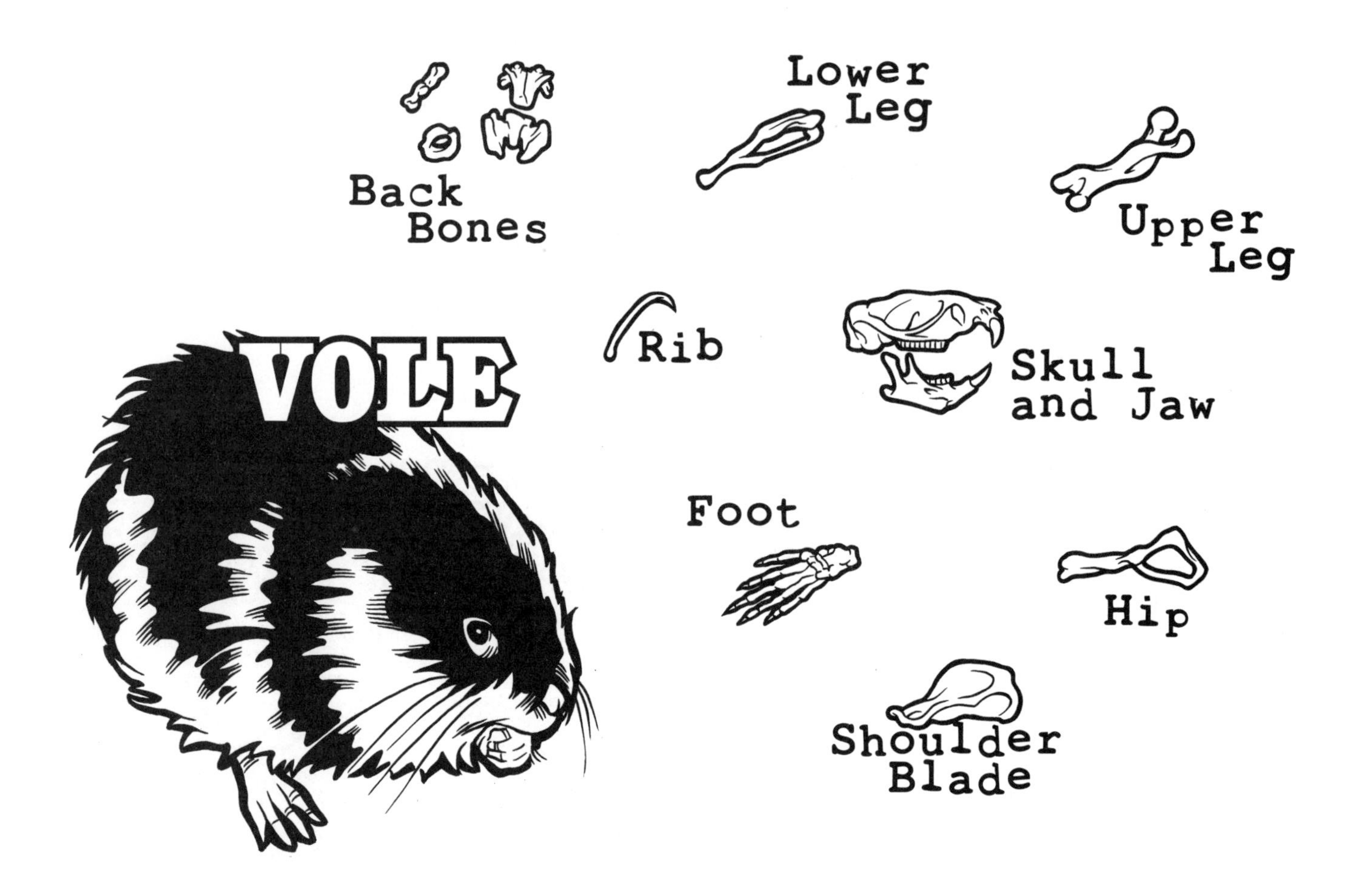
Back
Bones
Lower
Leg
Upper
Leg
Rib
VOLE
Skull
and Jaw
Foot
Hip
Shoulder
Blade

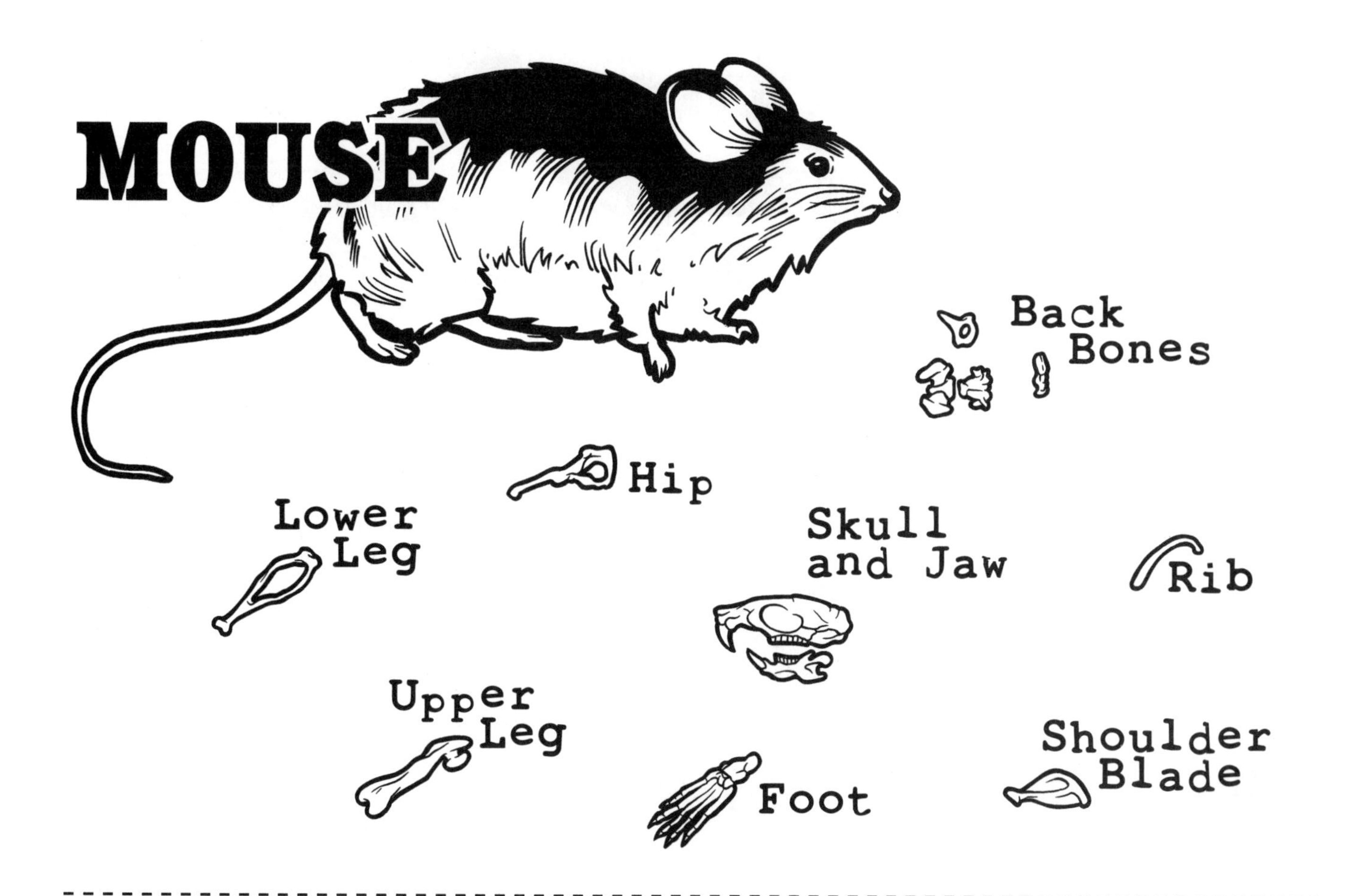
MOUSE
Back Bones
Hip
Lower Leg
Skull and Jaw
Rib
Upper Leg
Shoulder Blade
Foot

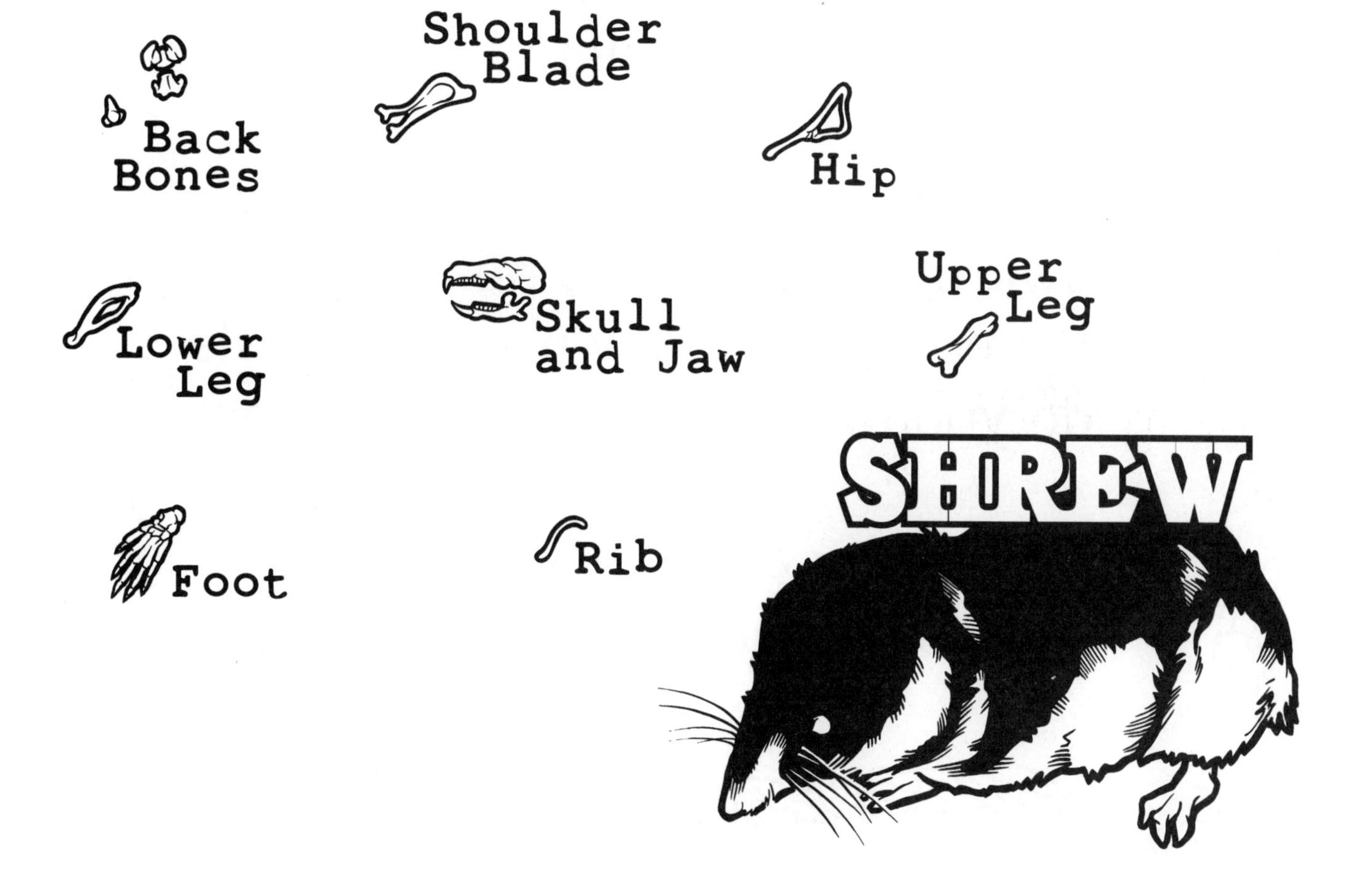
Shoulder Blade
Back Bones
Hip
Upper Leg
Skull and Jaw
Lower Leg
SHREW
Foot
Rib

Nocturnal Hunter

Connecting Learning

1. How were the owl pellets formed?

2. What adaptations help the owl catch its prey?

3. What is the average number of prey found within a pellet?

4. Studies have found that a typical owl regurgitates two pellets per day. Based on this, how many animals do you think an owl eats each day?

5. How many animals would an owl eat in one year?

Nocturnal Hunter

CONNECTING LEARNING

Connecting Learning

6. If an owl couple had four young owlets that must be fed for seven weeks until they mature, how much would the rodent population be reduced in this time?

7. What evidence did you find in the pellets that would tell you about the diet of the preys?

8. What "chain" can you connect between the sun and an owl using the evidence you found in the pellet?

9. What are you wondering now?

Here's Looking at You

Topic
Adaptations

Key Questions
1. How are owls adapted to help them hunt at night?
2. How do humans' field of vision, range of motion, and range of vision compare to that of an owl?

Learning Goals
Students will:
- learn how owls' eyes are adapted to help them hunt at night, and
- determine their range of vision using angle measurements and compare it to that of an owl.

Guiding Documents
Project 2061 Benchmark
- *One way to make sense of something is to think how it is like something more familiar.*

NRC Standards
- *Ask a question about objects, organisms, and events in the environment.*
- *Plan and conduct a simple investigation.*
- *Employ simple equipment and tools to gather data and extend the senses.*

*Common Core State Standards for Mathematics**
- *Model with mathematics. (MP4)*
- *Attend to precision. (MP6)*
- *Geometric measurement: understand concepts of angle and measure angles. (4.MD)*

Math
Measurement
angle

Science
Life science
adaptations

Integrated Processes
Observing
Collecting and recording data
Comparing and contrasting
Predicting

Materials
For each pair of students:
string, 60-70 cm
masking tape
clear protractor (see *Management 3)*
scissors
crayons or colored pencils

For each student:
Here's Looking at You rubber band book
four paper protractors
student page

Background Information

Owls are birds of prey that are especially adapted for hunting at night. Their large eyes enable them to see well under low light conditions. The structure of the eye itself allows owls to gather more light than diurnal birds (those active during the day). They have large pupils that allow more light to fall on the retina. In addition to the large pupil, they have many more rod cells, or light detecting cells, than the color detecting cone cells. Owls therefore can see in dim light, but have poor color vision, and are most likely colorblind.

The eyes of an owl face forward in a flat, broad feathered "facial disk" not found in other birds. In this way, both eyes see the same object at the same time, providing the bird with binocular vision. Binocular vision enables an owl to judge distances between objects, increasing the bird's ability to maneuver in a crowded environment.

Owls' eyes do not have any muscles that allow them to move. Facing forward, the field of vision for an owl is about 110 degrees, with about 70 degrees being binocular vision. Since owls' eyes cannot move in their sockets, an owl must turn its head to follow a moving object. A long, flexible neck aids this motion. The owl's neck is not always apparent because feathers and the owl's posture hide it. An owl's neck has more than twice as many vertebrae as a human's. It can turn its head 270 degrees—a range of motion much larger than that of humans.

Key Vocabulary

Field of vision: the entire area that a person or animal can see without moving their eyes, measured in degrees

Range of motion: the number of degrees the head can be turned to the right and to the left

Range of vision: the sum of the field of vision and the range of motion—the total number of degrees that a person/animal can see by turning the head

Management

1. Group the students into pairs. One student will be the observer and the other student will be the measurer. Then, the partners will switch roles.
2. It is assumed that students have had previous experience with using protractors to measure angles.
3. Copy the page of protractors onto both transparency film and white copy paper. Each pair of students needs one clear protractor, and each student needs four protractors on white paper.
4. To find their field of vision, students will measure the angle from zero degrees (looking straight forward) to the extent of their peripheral vision on the right and on the left. Students may not move their heads or their eyes when finding this measurement.

Procedure

Field of Vision

1. Ask the *Key Questions* and introduce the *Key Vocabulary*. Distribute the rubber band book and read through it as a class. Discuss how the *Key Vocabulary* relates to the information in the rubber band book.
2. Distribute the student page and colored pencils. Remind students of what they just read and direct their attention to the protractor that shows the owl's field of vision. Discuss how the dashed lines show a total of 110 degrees—55 degrees on each side of the zero. Have the students color in the region between the two dashed lines that represents the owl's total field of vision. Have them articulate what the field of vision represents. [what the owl can see without moving its head]
3. Discuss whether students think their field of vision is likely to be greater than an owl's, less than an owl's, or about the same as an owl's.
4. Give pairs of students one meter of string, masking tape, and a transparent protractor. Have them place a piece of masking perpendicular to the edge of a table and draw a line down the center of the tape. Show them how to tape the string to the edge of the table as illustrated.

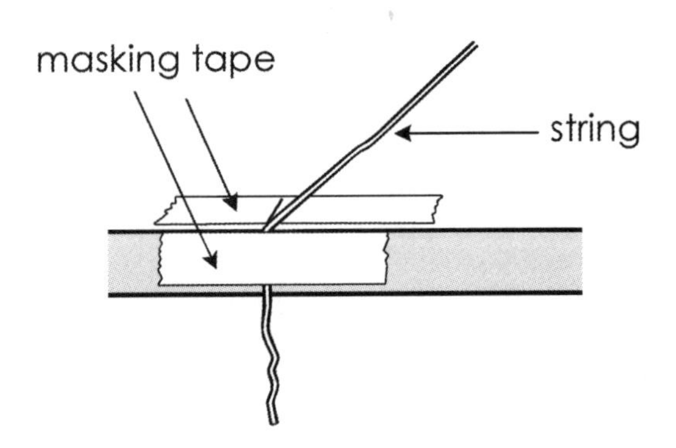

5. Explain the two roles—observer and measurer—and have students decide who will begin in each role. Ask the observer in each pair to kneel at the edge of the table and to align the bridge of his or her nose to the center of the tape. Tell the observers to look straight ahead and explain that they must not move their eyes. You may wish to suggest that they focus on some object directly in front of them to be sure that they are not turning their eyes.
6. Instruct the second student, the measurer, to tie a colored pencil or crayon to the end of the string. Have them hold it directly in front of the first student's nose and gradually move it around the observer's head toward the right, keeping the string taut at all times. Instruct the measurers to stop at the point where the observer can no longer see the colored pencil. Have the measurer hold the colored pencil and string in place.

7. Have the observer align the center of the transparent protractor with the line on the tape and the edge of the desk while the measurer holds the string in place. Have the students read the angle measure where the string hits. Ask the students to record the measurement on the student page on the first protractor labeled *Me*.

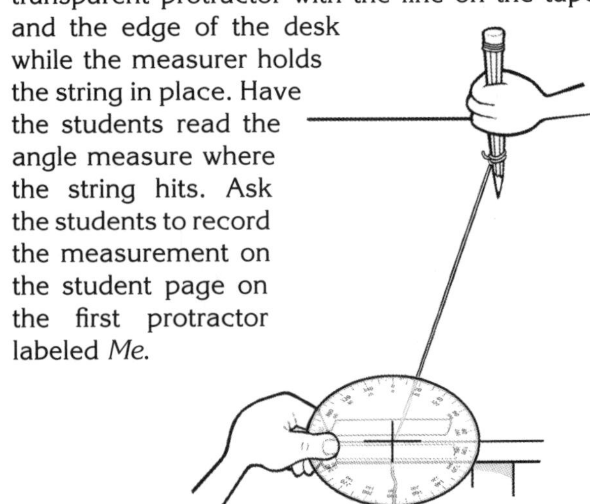

8. Have the students repeat the same procedure for the observer's left view.
9. Tell students to trade positions and repeat the process so that each student has determined his/her own total field of vision.

10. Ask students to compare the field of vision of an owl with their own. Invite students to share what they discovered and compare their results with their predictions.

Range of Motion

1. Have students use the information from the rubber band book to complete the protractor showing the owl's range of motion. They should color it to show a total of 270 degrees—135° to the left and 135° to the right.
2. Instruct students to use the same strings as before to measure how far to the right and to the left they can turn their heads. This time, the observer will align his/her head with tape again and then turn his or her head as far to the right as possible. With the observer's head turned as far right as possible, have the measurer move the string to align with the angle of the observer's nose and, using the clear protractor, determine the angle measurement. Have students record this information on the student page. Repeat this procedure with the students turning their heads as far left as possible.
3. Have students exchange roles and repeat the procedure so that all students have measured to determine their total range of motion.
4. Ask students to compare their range of motion with that of an owl. [an owl has a greater range of motion even though their field of vision is less than that of a human]

Range of Vision

1. Give each student a page of protractors on white copy paper. Have them label each of the protractors as follows: Field of Vision—Owl, Range of Motion—Owl, Field of Vision—Me, Range of Motion—Me.
2. Instruct them to use the information from the student page to draw and color the field of vision and range of motion for the owl and for them on the appropriate protractors.
3. Once students have colored the protractors, have them cut out the colored wedges. (The labels should be in the parts that are colored. If not, have students re-write the labels so that they will be on the colored wedges.)

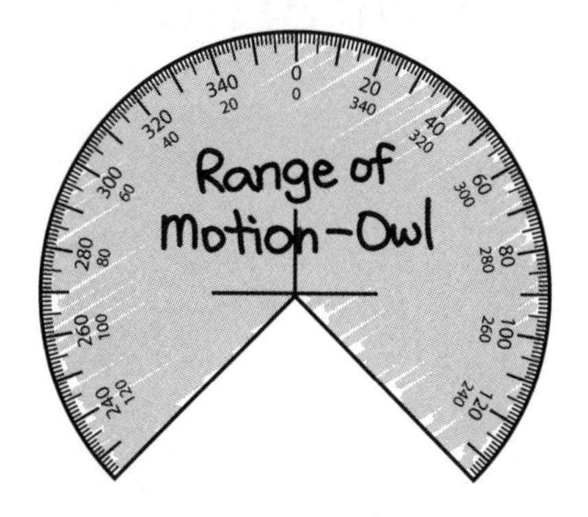

4. Instruct students to tape the two wedges that represent their field of vision and range of motion together along an edge. Ask them to identify what this represents. [their range of vision]
5. Have them complete the bottom portion of the student page by filling in the measurements to quantify their range of vision.
6. Repeat this process with the wedges that represent the owl's field of vision and range of motion. Compare the two representations and how many degrees the students can see compared to an owl.

Connecting Learning

1. What is your field of vision?
2. How does your field of vision compare to that of an owl? (It should be greater.)
3. What is your range of motion?
4. How does your range of motion compare to that of an owl? [It is quite a bit less.]
5. How did we find the total range of vision? [added the field of vision and the range of motion]
6. What is the total range of vision for an owl? [380°]
7. How does your total range of vision compare to that of an owl's? (It should be less.) How do you know? [The numeric measurement as well as the visual representation both show that the owl can see more than 360°. I can see less than 360°.]
8. How do you think this great range of vision helps owls? [Owls are predators; their range of vision allows them to see their prey, even when it is behind them.]
9. Do you think all birds have a similar range of vision? Explain. Do you think that birds that eat seeds have as large a range of vision as the owl? How could you find out?
10. What are you wondering now?

Here's Looking at You

Key Question

How does your range of vision compare to that of an owl?

Learning Goal

Students will:

use angle measurements to compare their range of vision with that of an owl.

Here's Looking at You

Owls are birds of prey. They are especially adapted for hunting at night.

1

The eyes of owls face forward. Both eyes see the same object at the same time. This lets them judge distances between objects. This is very important when hunting prey. They also have very large eyes with lots of light-detecting cells. This lets them see well in low light.

2

In fact, owls' eyes are so large, there is no room for muscles to move the eyes. They can only see about 110 degrees (55° to the left and 55° to the right) when they are looking forward. This is called their *field of vision*.

3

Since they can't move their eyes, owls must turn their heads to follow moving objects. Fortunately, they have long, flexible necks. They are able to turn their heads almost completely around. Their *range of motion* is 270 degrees—135° to the left, and 135° to the right.

4

CONNECTING LEARNING

Connecting Learning

1. What is your field of vision?
2. How does your field of vision compare to that of an owl?
3. What is your range of motion?
4. How does your range of motion compare to that of an owl?
5. How did we find the total range of vision?
6. What is the total range of vision for an owl?

CONNECTING LEARNING

Connecting Learning

7. How does your total range of vision compare to that of an owl's? How do you know?

8. How do you think this great range of vision helps owls?

9. Do you think all birds have a similar range of vision? Explain. Do you think that birds that eat seeds have as large a range of vision as the owl? How could you find out?

10. What are you wondering now?

Topic
Animal adaptations

Key Question
How can sea turtle eggs withstand a drop of 80 cm when being laid?

Learning Goals
Students will:
- read about the habits of nesting sea turtles,
- design a method to protect an egg dropped from 80 cm, and
- relate the activity to the adaptations of sea turtles and their eggs.

Guiding Documents
Project 2061 Benchmark
- *For any particular environment, some kinds of plants and animals survive well, some survive less well, and some cannot survive at all.*

NRC Standard
- *Each plant or animal has different structures that serve different functions in growth, survival, and reproduction. For example, humans have distinct body structures for walking, holding, seeing, and talking.*

Math
Measurement
  linear
  mass

Science
Life science
  adaptations

Integrated Processes
Observing
Comparing and contrasting
Collecting and recording data
Applying

Materials
For each student:
  eggs
  zipper-type plastic bags, pint size

For the class:
  balances
  masses
  metric rulers
  meter stick
  empty copy paper box

Background Information
Most male sea turtles never return to land after hatching. The females return to land to lay their eggs. A female will usually come ashore during the night. She crawls above the high tide mark where she digs a nest. The female first uses her front flippers to dig a body pit. Once the body pit is finished, the female nestles down in it and uses her hind flippers to dig an egg cavity. This egg cavity can be 80 centimeters deep.

A female will lay 50-200 eggs in about an hour. The eggs are about the size and shape of table tennis balls. They are leathery and covered in mucus. This helps prevent their breakage when they fall into the egg cavity.

Once the female is finished laying her eggs, she covers up the egg cavity. The sand that covers the eggs helps keep the eggs from drying out, helps keep them at the proper temperature, and helps protect them from predators.

Management
1. Students will need to work on this assignment at home. It may be best to assign it for over a weekend.
2. Set up a testing area where all eggs can be tested. Place the bottom part of the copy paper box against a wall. Place a meter stick inside the box, zero end in the bottom of the box, and tape the meter stick to the wall. Students will use this setup for dropping their eggs.

3. Balances (item number 1917), gram masses (item number 1923), metric rulers (item number 1909), and meter sticks (item number 1908) are available from AIMS.

Procedure

1. Ask the *Key Question* and state the *Learning Goals.*
2. Distribute the student pages. Let the students read the passage about female sea turtles and their eggs. Discuss the passage.
3. Inform the students as to what day they will have the egg drop. Tell them that you will send an egg home with them in a plastic bag. They are to determine how to protect it for the drop. Go over the rules for protecting their eggs.
4. Allow time for students to determine the masses of their eggs. If any eggs break before the egg drop, students will need to find the mass of the replacement eggs.
5. Have a piece of paper available so that students who have met all the qualifications can sign up for the egg drop.
6. Make sure that students hold the bottom of the protected egg even with the 80-centimeter mark in order to drop it.
7. Let each student unpack his/her egg to check whether it survived the drop.
8. Discuss the results.

Connecting Learning

1. How do sea turtle eggs differ from chicken eggs?
2. How was our experience similar to what the sea turtle's eggs do?
3. How is the sea turtle adapted for laying eggs?
4. How are the eggs adapted?
5. The hatchlings have a temporary egg tooth that helps them break out of the shell. What other adaptations do you think the hatchlings need? [a way to crawl out of the deep sand]
6. What are you wondering now?

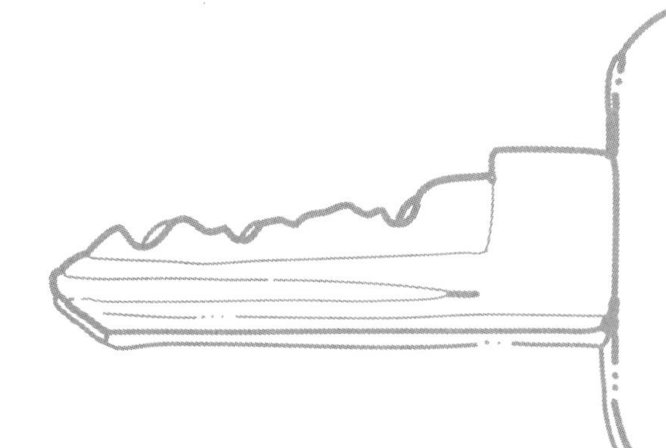

Key Question

How can sea turtle eggs withstand a drop of 80 cm when being laid?

Learning Goals

Students will:

- read about the habits of nesting sea turtles,
- design a method to protect an egg dropped from 80 cm, and
- relate the activity to the adaptations of sea turtles and their eggs.

An adaptation is something—a behavior, a structure, or a trait—that helps an organism survive.

Most male sea turtles never return to land after hatching. The females return to land to lay their eggs. A female will usually come ashore during the night. She crawls above the high tide mark where she digs a nest. The female first uses her front flippers to dig a body pit. Once the body pit is finished, the female nestles down in it and uses her hind flippers to dig an egg cavity. This egg cavity can be 80 centimeters deep.

A female will lay 50-200 eggs in about an hour. The eggs are about the size and shape of table tennis balls. They are leathery and covered in mucus. This helps prevent them from breaking when they fall into the egg cavity.

Once the female is finished laying her eggs, she covers up the egg cavity with sand.

The sand:

- helps keep the eggs from drying out,
- helps keep them at the proper temperature, and
- helps protect them from predators.

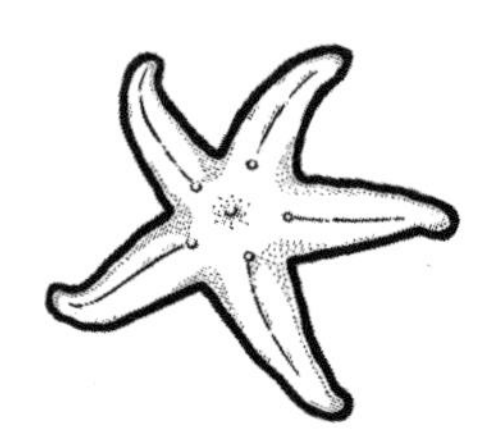

From this reading passage, what adaptations do the sea turtles have that help them lay eggs?

How are the eggs adapted for survival?

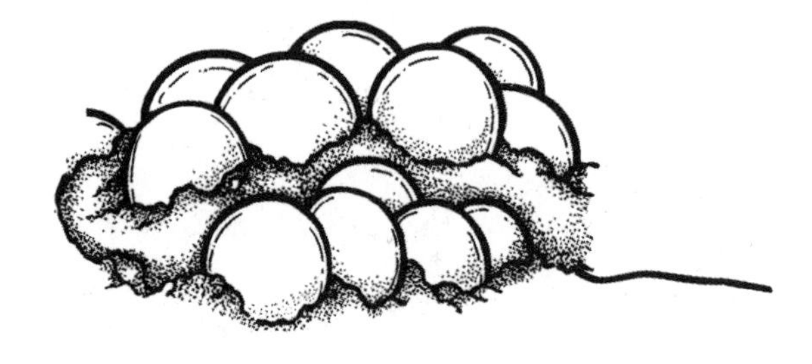

Your task is to find a way to protect an egg so that it can withstand an 80-centimeter drop.

Here are the rules:

1. You will use a chicken egg.
2. You can protect it with materials of your choice. However, you cannot add to its mass more than 100 grams.
3. The final "egg" cannot take up more space than 10 cm x 10 cm x 10 cm.

Data:

Mass of egg: ______________________

Mass of egg and protective material: ______________________

Difference in the mass of the egg and the protected egg: ______________________

Dimensions of protected egg: ______________________

Testing your egg:

Did you meet all the rules for the egg drop? If so, you can place your name on the list to test your protected egg.

Results:

Did your egg survive?

How is this like the sea turtle's egg?

How is it different?

Other Adaptations of Sea Turtles

Sea turtles get rid of the salt from the ocean water through a salt gland in their eyes. The excess salt comes out of their eyes. Observers say that it looks like the turtle is crying.

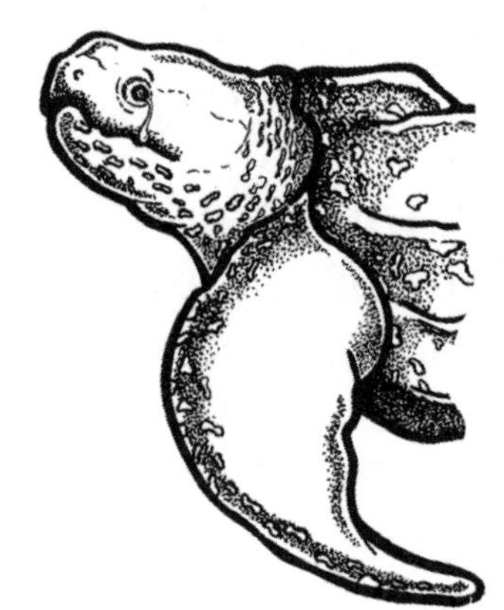

Sea turtles have front flippers for paddling.

Sea turtles have a shell for protection. Even though the shell is heavy and slows them down on land, in the water the turtle is streamlined and swims very well.

Sea turtles can't pull their heads and legs into their shells like land tortoises. This would alter their shape and hinder their swimming ability.

Female sea turtles return to their birthplace to lay eggs.

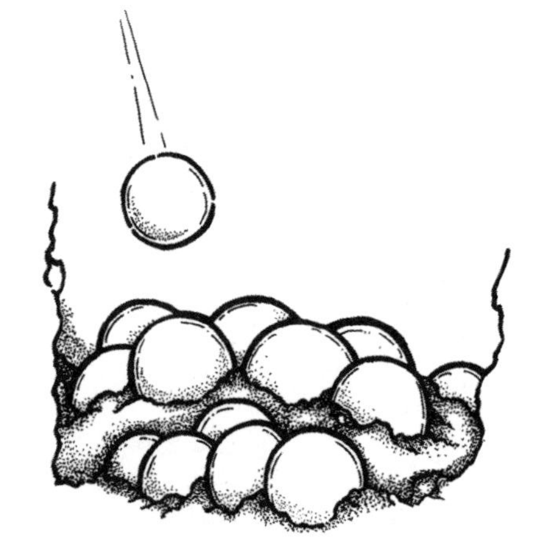

CONNECTING LEARNING

Connecting Learning

1. How do sea turtle eggs differ from chicken eggs?

2. How was our experience similar to what the sea turtle's eggs do?

3. How is the sea turtle adapted for laying eggs?

4. How are the eggs adapted?

5. The hatchlings have a temporary egg tooth that helps them break out of the shell. What other adaptations do you think the hatchlings need?

6. What are you wondering now?

Deep Divers

Topic
Animal adaptations

Key Question
How are whales, dolphins, and other marine mammals able to swim so deep?

Learning Goals
Students will:
- read about the research work of marine biologist, Dr. Terrie Williams;
- use a balloon and paper clips to make a simple model of a marine mammal; and
- use the model to observe deep diving.

Guiding Documents
Project 2061 Benchmarks
- *A model of something is different from the real thing but can be used to learn something about the real thing.*
- *A great variety of kinds of living things can be sorted into groups in many ways using various features to decide which things belong to which group.*

NRC Standard
- *Each plant or animal has different structures that serve different functions in growth, survival, and reproduction.*

Science
Life science
  adaptations

Materials
For each group:
  recycled two-liter bottle with cap
  small balloon
  15 paper clips
  plastic cup, 16 oz (at least 6 inches high)
  paper towels

Background Information
Humans use a lot of energy swimming. To swim underwater and touch the bottom at the deep end of the pool, we take a deep breath and then move our arms and legs to stroke our way down. We don't dive very deep before our lungs and muscles ache for more oxygen. Soon we rise to the surface for a big gulp of air.

Whales, seals, and dolphins have to breathe air like us, but they spend much time underwater. When marine mammals dive, they hold their breath. For shallow dives, they use their tails and flippers to move. If they used tails and flippers to propel themselves during deep dives, they would not have enough energy and oxygen to dive very deep. But these animals are known to reach depths as great as 400 meters (over 1200 feet), which is far greater than human divers can reach. How are they able to do this?

In the late 1990s, Dr. Terrie Williams, a marine biologist at the University of California, Santa Cruz, headed a team of marine scientists that attached video cameras to several species of deep-diving marine mammals to find out how these animals are able to go to such depths. They found that deep-diving mammals do not stroke with tails and flippers to such depths. Instead, they sink! Sinking requires much less energy and oxygen use than active swimming. They steer themselves with body motions, but their speed comes from changing their bodies so that they will sink.

As the animals descend, there is a decrease in lung volume as the lungs gradually collapse due to increasing water pressure. For a bottlenose dolphin, complete collapse of the lung occurs when a depth of approximately 70 meters is reached.

The collapse of the lung decreases the volume of the dolphin, which causes it to sink. As the dolphin sinks, it uses its flippers and tail to steer its body. Such sinking avoids the energy costs of stroking with flippers and tail. The lung collapse also keeps high-pressure gas from getting into their bloodstreams, so they don't suffer from the bends when ascending. The bends is a painful and sometimes fatal condition that occurs when a human diver comes up too rapidly.

Dr. Williams and her team attached video cameras and data loggers to each of four different species of marine mammals: three adult Weddel seals diving from an isolated ice hole in Antarctica; a young elephant seal diving in Monterey Bay, California; an adult bottlenose dolphin diving offshore of San Diego, California; and an adult blue whale diving offshore of northern California.

The video cameras strapped to each animal recorded the motion of the flippers and tail, and the data loggers recorded the duration of the dive and the maximum depth.

Management

1. This activity requires small balloons. Do not use a round balloon that inflates to over 7 inches or an uninflated sausage balloon longer than 4 inches.
2. Blow a single puff into the balloon. The sides of the balloon should not be near fully inflated. Stretch and knot the neck of the balloon.

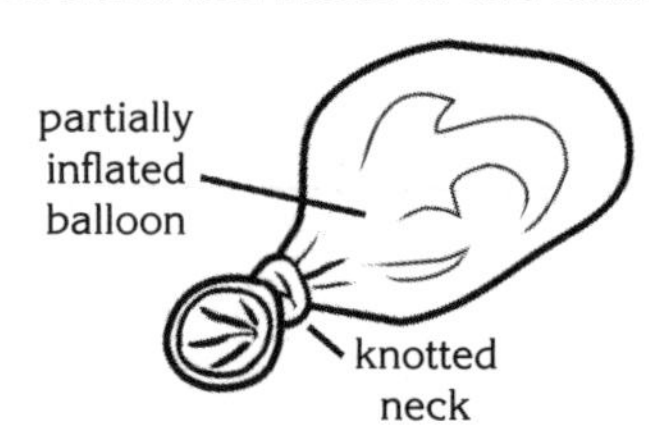

3. Thread the knotted neck of the balloon *through* the smaller end of a paper clip. Be sure the outside end of the clip is pointing up. This is the working model of a deep-diving whale or dolphin.

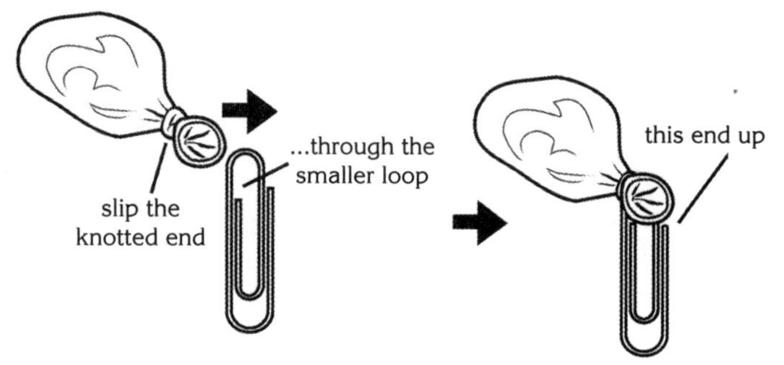

4. Make one balloon and paper clip model for each group of students.

Procedure

1. Organize the students into groups of two or three.
2. Display the first page using a projection device and use the page to introduce the activity.
3. Hand out a copy of the reading passage to every student. Have the students read the page. Hold a whole-class discussion about the content.
4. Distribute a balloon with attached paper clip and 15 additional paper clips to every group.
5. Make sure that each group has a 16-ounce plastic cup and paper towels. Assign one student in each group to fill the cup with water.
6. Distribute the *Deep Divers* student page to every student.
7. Direct one student in each group to hang one of the 15 paper clips on the paper clip attached to the balloon and to float the balloon and clips in the plastic cup of water.
8. Tell the students to take turns hanging paper clips on the clip attached to the balloon until the tip of the balloon is just above the surface of the water.

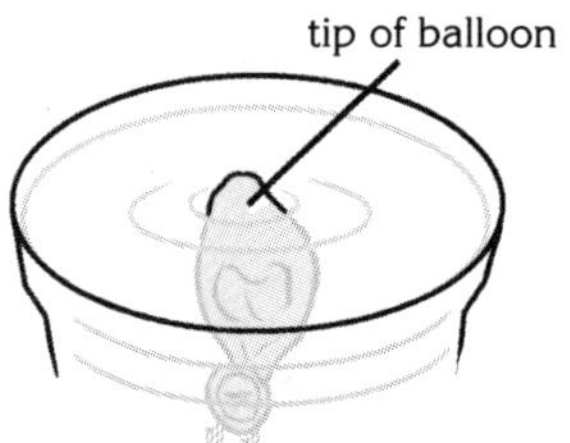

   Inform the students that the balloon and paper clips model the floating behavior of a whale or dolphin.
9. Distribute a two-liter bottle and cap to every group. Assign one student in each group to fill the bottle with cold water.
10. Demonstrate for the students how to carefully lower the balloon and paper clips, paper clips first, into the neck of the two-liter bottle. Show them how to use the eraser end of a pencil to push down on the top of the balloon so that the model is not trapped in the neck of the bottle.
11. Direct one student in each group to use the water in the plastic cup to completely fill the two-liter bottle and then tightly cap it.
12. Tell the students in each group to take turns squeezing the sides of the bottle and observing the behavior of the "marine mammal." Challenge the students to observe that the balloon gets slightly smaller when the sides are squeezed and the model sinks. It returns to normal size when the model returns to the surface.
13. Have the students describe how the model is like a real whale or dolphin and how it is different.

Connecting Learning

1. Does it take more energy to walk 100 meters or run 100 meters? How do you know?
2. When you walk, do you think that it would take more or less energy to swing your arms in a big arc than to not swing them? Why do you say this?
3. How do dolphins and whales move through the water? Do you think this takes energy? Explain.
4. Why would dolphins and whales need to conserve energy when they dive deep? [so they can stay down longer]
5. How does our model show they do that?
6. What happens to our bodies when we breathe deeply and hold our breath? [our lungs fill up and our chest gets larger]
7. What happens to the lungs of a dolphin or whale when they dive deep? [They collapse.]
8. How is the model (balloon and paper clips) like a real whale or porpoise? [model sinks without swimming, balloon models the lungs of a marine mammal, etc.]
9. How is the model different from a real whale or porpoise? [The balloon and clips only model the lungs of a marine mammal. None of the other systems, like the heart, tail, flippers, etc., are modeled.]
10. What are you wondering now?

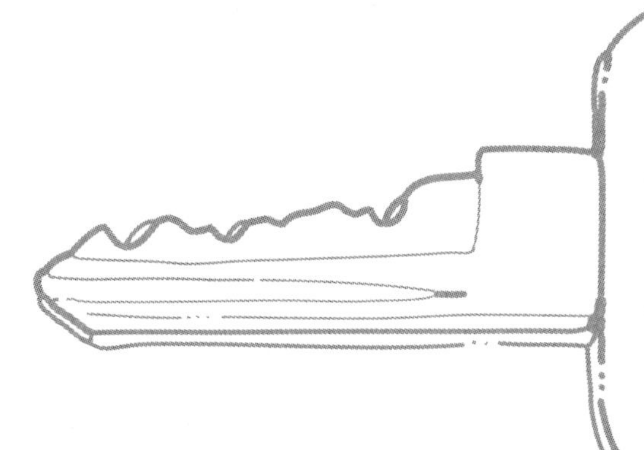

Key Question

How are whales, dolphins, and other marine mammals able to swim so deep?

Learning Goals

Students will:

- read about the research work of marine biologist, Dr. Terrie Williams;
- use a balloon and paper clips to make a simple model of a marine mammal; and
- use the model to observe deep diving.

Deep Divers

How Do Whales, Seals, and Dolphins Dive to Great Depths?

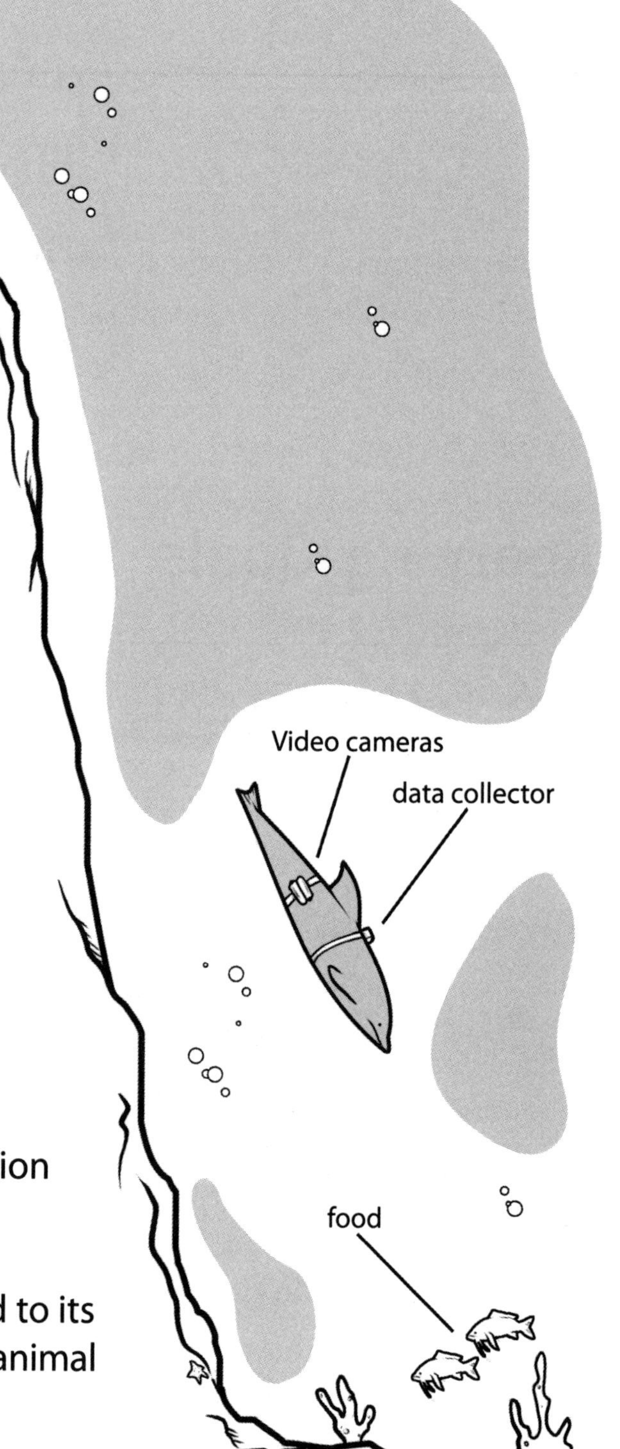

A **marine biologist** is a scientist who studies life in the oceans and the seas.

A **mammal** is an animal with a backbone that has hair or fur and breathes air using lungs.

Marine mammals, like whales, seals, and dolphins, are animals that live in the seas or oceans.

Dr. Terrie Williams, a marine biologist at the University of California, Santa Cruz, led a team of scientists that studied how marine mammals, like whales, seals and dolphins, dive deep.

The scientists couldn't figure out how the animals were able to breathe in enough air for them to dive as deep as they could.

The scientists strapped video cameras to the animals' backs. The cameras recorded the motion of their flippers and tails.

Each animal also had a data collector strapped to its back. This instrument recorded how long the animal held its breath and how deep the animal went.

Some marine mammals, like the **elephant seal**, can dive as deep as 400 meters!

Deep Divers

How do Whales, Seals, and Dolphins Dive to Great Depths?

People use much more energy to swim than to walk or run.

To swim to the bottom of a pool, we take a deep breath and move our arms and legs to take us down. We don't dive very deep before our lungs and muscles ache for more air. Soon, we rise to the surface for a big gulp of air.

Whales, seals, and dolphins live in the water. They have to breathe air like us. But they spend much more time underwater looking for food. When they dive, they also have to hold their breath. For shallow dives, they use their tails and flippers to move. If they used tails and flippers to swim deep, they would not have enough energy and oxygen to dive very deep. But these animals are known to reach depths as great as 400 meters. That's the length of four football fields! How are they able to do this?

In the late 1990s, Dr. Terrie Williams led a team of scientists that wanted to study how whales and dolphins dive. They attached cameras to the backs of these animals. They found that the animals do not use tails and flippers to swim down. Instead, they sink! Sinking requires much less energy and air. They do use flippers and tails to steer. But their speed comes from changing their bodies so that they will sink.

As the animals dive, their lungs slowly get smaller. This is caused by the increasing weight of the water on the animals. You can feel this weight on your ears when you dive. For a dolphin, the lung becomes flat near a depth of 70 meters.

Sinking allows whales and dolphins to dive deep on a small supply of air. To get back to the surface, they use their flippers and tails to swim up.

How do penguins dive deep? Dr. Williams doesn't know. As birds, they don't breathe the same way as whales and dolphins. A future scientist will probably discover how penguins dive deep.

Deep Divers

1. Add paper clips to the balloon until the tip of the balloon is just above the water level.

tip of balloon

2. Put your model into the bottle. Screw the lid on tightly.

3. Press the sides of the bottle and describe what you observe.

4. How is your model like a real whale or dolphin?

5. How is your model not like a real whale or dolphin?

CONNECTING LEARNING

Connecting Learning

1. Does it take more energy to walk 100 meters or run 100 meters? How do you know?

2. When you walk, do you think that it would take more or less energy to swing your arms in a big arc than to not swing them? Why do you say this?

3. How do dolphins and whales move through the water? Do you think this takes energy? Explain.

4. Why would dolphins and whales need to conserve energy when they dive deep?

5. How does our model show they do that?

Connecting Learning

6. What happens to our bodies when we breathe deeply and hold our breath?

7. What happens to the lungs of a dolphin or whale when they dive deep?

8. How is the model (balloon and paper clips) like a real whale or porpoise?

9. How is the model different from a real whale or porpoise?

10. What are you wondering now?

Topic
Adaptations

Key Question
What adaptations do turkey vultures have to meet their needs?

Learning Goals
Students will:
- investigate the adaptations that help turkey vultures meet their needs, and
- determine how wild animals in their area are adapted to meet their needs.

Guiding Documents
Project 2061 Benchmark
- *For any particular environment, some kinds of plants and animals survive well, some survive less well, and some cannot survive at all.*

NRC Standard
- *An organism's behavior evolves through adaptation to its environment. How a species moves, obtains food, reproduces, and responds to danger are based in the species' evolutionary history.*

Science
Life science
  adaptations
  habitats
  animal behavior

Integrated Processes
Observing
Comparing and contrasting
Collecting and recording data
Applying

Materials
Turkey Vultures rubber band book
#19 rubber bands
Student pages
Internet access
Reference materials on animals

Background Information
Both plants and animals have adaptations that help them meet their basic needs. These adaptations are the inherited traits or characteristics that help the organism get the food, air, water, and shelter it needs in its habitat. Turkey vultures, for example, have two adaptations that help them find their food (carrion) while soaring high in the sky: keen eyesight and a highly developed sense of smell. (This latter adaptation is rare among birds.) In addition, turkey vultures have a unique digestive system that kills the bacteria and viruses that are present in rotting carcasses. These, and several other adaptations, help turkey vultures carve out an important niche as natures' "carrion disposals." Like turkey vultures, other wild animals have special adaptations that let them fill niches in their habitats.

Management
1. This activity has two parts. In the first part, students read the rubber band book on turkey vultures and answer questions about their adaptations. The second part has the students select a wild animal that lives in their area and list its needs and adaptations.
2. The second part of this activity can be done in several ways. If students know a lot about the wild animals they pick, they can answer the questions about their needs and adaptations directly. If they don't know much about their animals, they can use the Internet, encyclopedias, or library books to do some research.

Procedure
1. Have students read the rubber band book *Turkey Vultures.*
2. Ask the *Key Question:* "What adaptations do turkey vultures have to meet their needs?"
3. Distribute the first student page and have students answer the questions about the turkey vulture's adaptations.
4. Distribute the second student page. Have students brainstorm and list wild animals that live in the area. (These animals can include such things as insects, spiders, reptiles, and fish as well as birds and mammals.)

5. Have students choose one of the animals brainstormed and record some of the adaptations that animal has that help it meet its needs.
6. End with a time of class discussion where students share their animals, their animals' needs, and their animals' adaptations to meet those needs.

Connecting Learning

1. What adaptations help turkey vultures find their food? [They have keen eyesight and an excellent sense of smell.]
2. What adaptations help turkey vultures eat carrion? [Their digestive systems are adapted to kill the germs present in the rotting carcasses. Their lack of head feathers helps keep them clean as they eat.]
3. What needs does your animal have?
4. What adaptations does your animal have to meet its needs?
5. If the turkey vulture were to become extinct, what other animal could fill its niche? [The animal would have to be one that could find and eat carrion without getting sick and spreading germs.]
6. What are you wondering now?

Extensions

1. Have students brainstorm plants in the area and list their adaptations.
2. Have students write a report on their chosen animals.

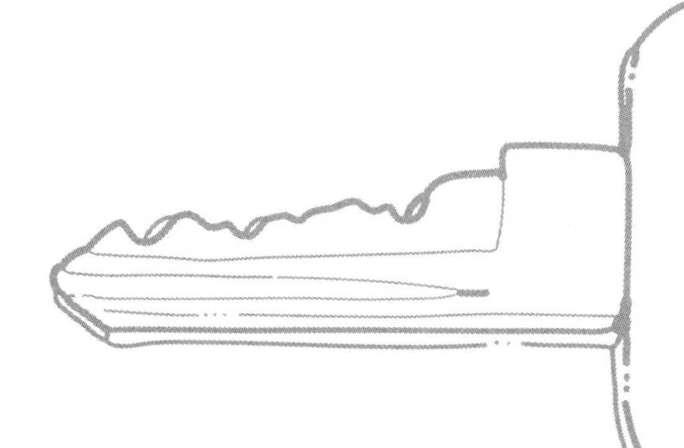

Key Question

What adaptations do turkey vultures have to meet their needs?

Learning Goals

Students will:

- investigate the adaptations that help turkey vultures meet their needs, and
- determine how wild animals in their area are adapted to meet their needs.

Turkey vultures are found in most parts of North and South America. Many people think they are ugly because of their featherless red heads. When flying, however, they are often mistaken for eagles.

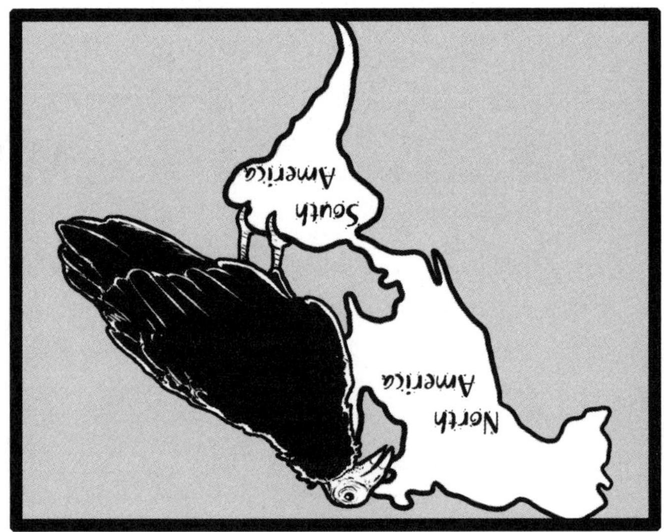

Turkey vultures are quite intelligent and social. They live together in large family groups called roosts. They also love to play. Sometimes turkey vultures become attached to humans. They follow them as they take daily walks.

The turkey vulture's excellent sense of smell is unusual for birds. Gas companies sometimes use this sense to find leaks in gas pipelines. They pump a smelly gas through the pipes. Then they wait to see where the turkey vultures begin to circle.

We should not see turkey vultures as ugly scavengers. We should appreciate them. They are amazing creatures that fill an important purpose in nature.

Turkey vultures are related to California condors. Like their larger cousins, they are scavengers. That means they eat dead animals (carrion). They spend hours gliding on breezes looking for food. They have two adaptations that help them find carrion from far away. They can see very well and have an excellent sense of smell.

3

4

Turkey vultures fill an important niche in their environments. They clean up dead animals. They can eat rotting meat that would make most other animals sick. Their digestive systems are adapted to kill harmful germs. Scientists are studying this unique digestive system to help them learn how to better fight germs.

5

The turkey vultures' lack of head feathers is another adaptation. There are no feathers to get messy as they eat. Although turkey vultures eat carrion, they are actually quite clean. They spend hours preening their feathers. If water is available, they will bathe and then spread their wings to dry.

The scientific name of the turkey vulture is *Cathartes aura*. That means, "cleansing breeze." The Cherokees call the turkey vulture "peace eagle." It looks like an eagle, but doesn't kill.

6

Animal Adaptations

Adaptations are inherited traits that help a plant or animal meet its needs. These needs are things like getting food, water, and air; finding shelter; protecting themselves; and reproducing.

The book on turkey vultures mentioned several adaptations that help these birds find their food. What are these adaptations?

How do they help the vultures find their food?

The book also mentioned some adaptations that help turkey vultures eat carrion. What are these adaptations?

How do they help the turkey vulture?

Animal Adaptations

Brainstorm a list of interesting animals that live in your area. (Don't forget things like insects and spiders.)

Pick one of these animals and list it here.

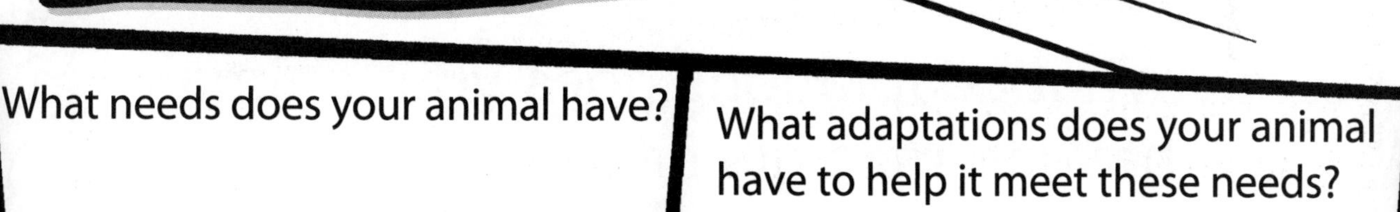

What needs does your animal have?

What adaptations does your animal have to help it meet these needs?

Share your findings with others.

CONNECTING LEARNING

Connecting Learning

1. What adaptations help turkey vultures find their food?
2. What adaptations help turkey vultures eat carrion?
3. What needs does your animal have?
4. What adaptations does your animal have to meet its needs?
5. If the turkey vulture were to become extinct, what other animal could fill its niche?
6. What are you wondering now?

Sometimes blending in helps animals catch a meal. Crab spiders hide in flowers. Their colors match the flowers. When an insect comes to the flower, the spider catches it.

All animals need to survive. Many have special characteristics that help them stay alive. We call these characteristics *adaptations*.

Think of some other animal defenses and list them here.

9

Eyespots help many bugs survive. These spots may make an animal look like something else. Sometimes they scare the bird or lizard away. Other times, they make the tail look like the head.

4

Monarch butterflies and ladybugs have a bad taste. When a bird eats one, it spits it out. Soon that bird learns not to eat monarchs or ladybugs.

Blending in can keep an animal safe. If it can't be seen, it can't be eaten. Walking sticks look like twigs on trees. Birds cannot spot them. Snowshoe hares have white fur in winter. This helps them blend in with the snow. In summer, they have brown fur. This helps them blend in with the dirt.

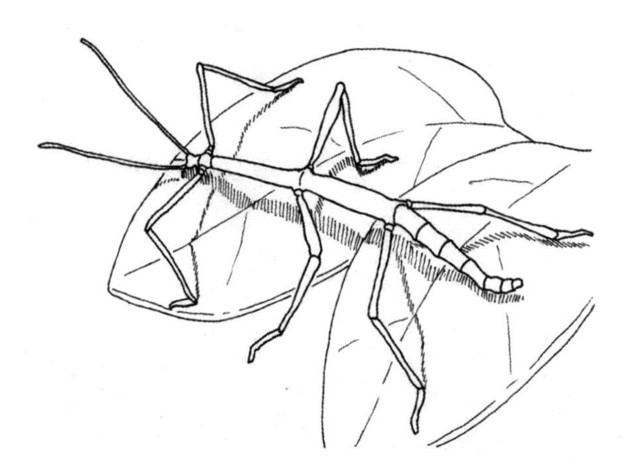

10

Some animals make a bad smell. This warns other animals to stay away. Skunks and stinkbugs use this defense.

3

The viceroy butterfly looks like the monarch. Birds are afraid to eat it. They do not want a bad tasting meal. But the viceroy does not have a bad taste. Its looks help it survive. This is called *mimicry*.

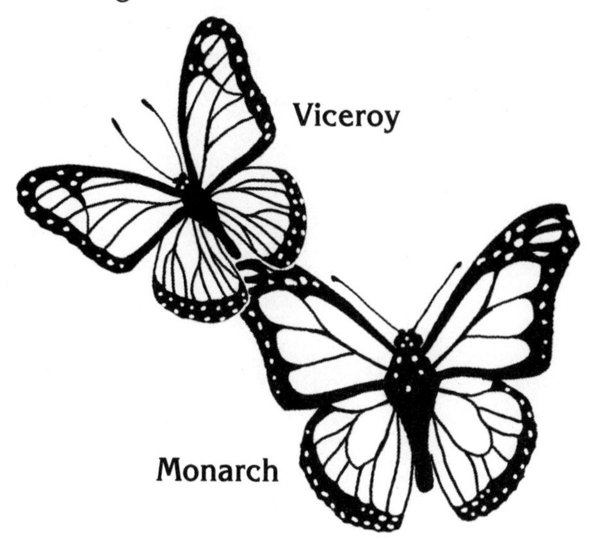

5

If an opossum is in danger, it plays dead. Many animals only attack live or moving animals. If the opossum lies still, it may be safe. Hognose snakes and snout beetles also play dead.

6

Some animals' defenses are their bodies. Porcupines have sharp quills. They point them at attackers.

Armadillos have hard shells. If attacked, some roll into a ball. This keeps them safe.

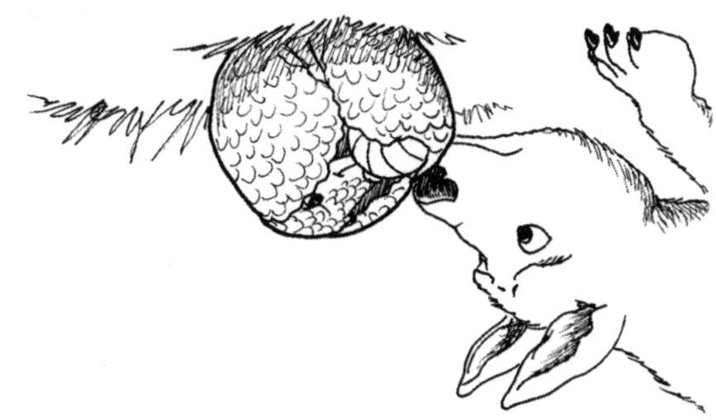

7

Some animals can grow new body parts. This helps them escape. If a cat catches a lizard by its tail, the tail will break off. Then the lizard is free. The tail will grow back in time. Some crabs can lose a claw to escape. Matchstick bugs can shed a leg if they are caught.

8

Defense by Design

Topic
Animal defenses

Key Question
How do animals defend themselves?

Learning Goals
Students will:
- identify ways in which animals defend themselves,
- construct an animal with specific defensive structures, and
- predict which defensive structures an animal has.

Guiding Documents
Project 2061 Benchmark
- *Models are very useful for communicating ideas about objects, events, and processes. When using a model to communicate about something, it is important to keep in mind how it is different from the thing being modeled.*

NRC Standard
- *Each plant or animal has different structures that serve different functions in growth, survival, and reproduction. For example, humans have distinct body structures for walking, holding, seeing, and talking.*

Science
Life science
 adaptations
  defense

Integrated Processes
Observing
Collecting and recording data
Classifying
Comparing and contrasting
Predicting

Materials
For each student:
 Animal Defenses rubber band book
 Mystery Defense Envelope (see *Management 1)*
 large marshmallow
 index card, 3" x 5" (see *Management 2)*

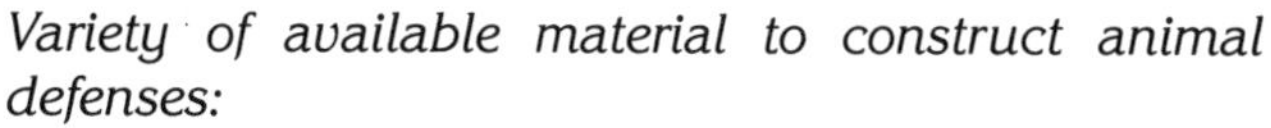

Variety of available material to construct animal defenses:
 toothpicks
 chenille stems
 construction paper
 drinking straws
 miniature marshmallows
 glue
 scissors

Background Information

Animals have adapted defenses to help them survive. These defenses can be behavioral or structural. Animals often feed in herds. When a predator attacks, the animals scatter and run in different directions. This confuses the predator and allows the animals to escape. Some animals never venture too far from their home in underground dens or thick vegetation and can quickly hide when danger approaches. Many animals have keen senses of sight, smell, and hearing so that they can detect danger and escape. Some animals have horns or antlers to fight off predators. Some animals rely on camouflage to blend in with their surroundings to hide from predators. A few animals are even poisonous or unpleasant tasting. Predators have learned to leave them alone. Some animals use chemicals that they spray from various parts of their bodies. A few animals rely on trickery and copy the defenses of other animals to protect themselves. Animal defenses are varied and each help in an animal's survival in an environment.

Management
1. To prepare the *Mystery Defense Envelopes,* cut apart two or three of the *Animal Defense Cards* and place them in a business-size envelope. Prepare an envelope for each student.
2. Each student will need an index card with an identifying number on it. Fold the cards in half so they will stand on their own. Write a number on the bottom half of each index card. The numbers will be placed beside each animal so that it will be easier to compare the students' data.

Procedure

1. Ask the *Key Question* and state the *Learning Goals.*
2. Distribute the *Animal Defenses* rubber band book. Read and discuss some of the ways in which animals defend themselves. Point out that even large animals like lions and elephants have defense structures.
3. Explain to the students that they will construct an imaginary animal that has specific defensive structures. Tell them that each animal's basic body will be a large marshmallow. They will decide what type of animal to make after they see what types of defense structures are in their envelope.
4. Distribute the *Mystery Defense Envelopes* and allow time for animal construction.
5. Guide the students in displaying the animals they created along with their numbered index cards. Tell them to put the *Animal Defense Cards* back inside their *Mystery Defense Envelopes* and place them beside their animals.
6. Distribute the *Animal Defense Safari Survey Sheet.* Explain to students that they will put a checkmark for each defensive structure they think the animal has. Tell them that they should make predictions for six animals.
7. Have the students complete the survey. When the students have completed the survey, direct them to take the cards out of the *Mystery Defense Envelopes* and place the cards beside the animals. Allow time for the students to check their predictions.

Connecting Learning

1. What are some ways in which animals defend themselves?
2. Why do you think animals have different types of structures for defense?
3. What are some ways claws and teeth can be used other than for defense?
4. How did building the model animal help you learn about animal defenses?
5. What are you wondering now?

Defense by Design

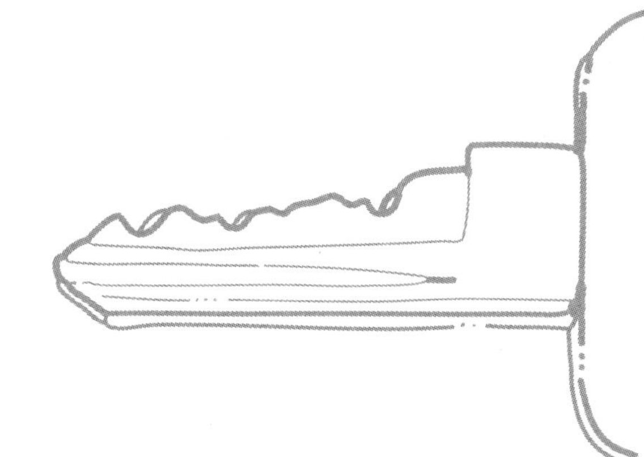

Key Question

How do animals defend themselves?

Learning Goals

Students will:

- identify ways in which animals defend themselves,
- construct an animal with specific defensive structures, and
- predict which defensive structures an animal has.

Animal Defenses

1

Animals have defenses. Defenses help to protect them from their enemies.

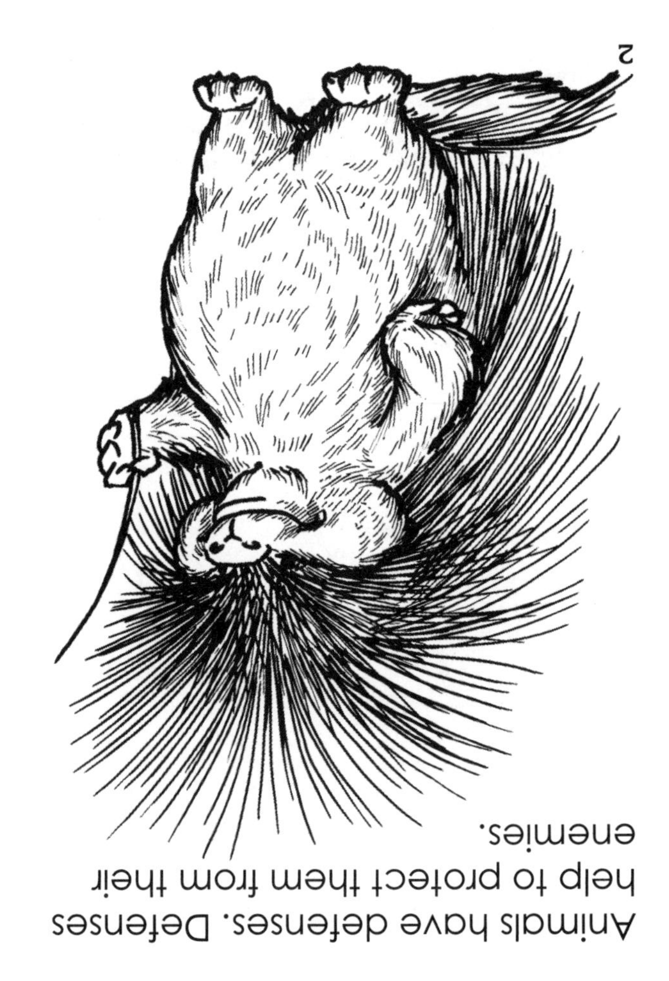

2

Other animals use poison or stings as a defense.

7

There are other animal defenses. Which ones can you think of?

Some animals use flight as a defense. They run, hop, swim, or fly away from a predator. This defense works if the animal is faster than its enemy.

3

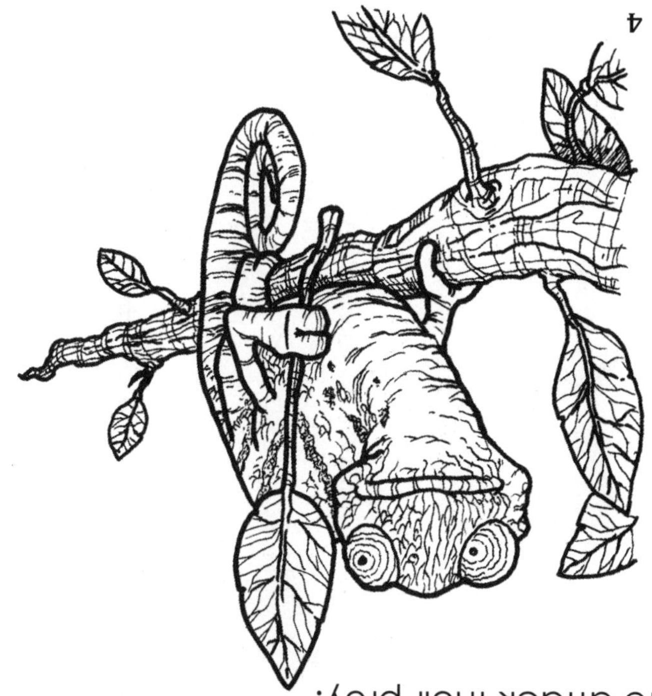

Some animals hide. Camouflage is their defense. Animals use camouflage to hide from their enemies. They also use camouflage to attack their prey.

Other animals use tricks as a defense. A puffer fish blows itself up to make it look bigger. Some animals ruffle their feathers or fur. This makes them look larger. Animals like the opossum and hognose snake play dead.

Animals like skunks give off a bad smell. They use a chemical defense. Porcupines use their physical structure for defense.

6

Defense by Design

Claws are an animal defense. They are on an animal's feet. The sharp claws can be used to fight and to catch and hold prey. Bears, owls, and cats use claws.

Quills and **spikes** are types of animal defenses. They are sharp and often protect the body of the animal. Porcupines, hedgehogs and lionfish use quills and spikes to protect themselves.

Horns and **antlers** are animal defenses. They are on the head of an animal. They are hard and can be used to fight. Deer, moose, goats, and water buffalo use horns and antlers as a defense.

Hard shells and **body coverings** are types of animal defenses. Animals use a hard outside covering for protection. Turtles, clams, and armadillos use hard shells and body coverings for defense.

Teeth are also an animal defense. They can be used for defense and to catch prey. Mice, wolves, squirrels, and woodchucks use their teeth for defense.

Tails are an animal defense. Tails can be used to hit enemies or knock them over. Some animal tails can break off so the animal can escape. Kangaroos, lizards, and alligators use tails for defense.

Defense by Design

Animal Defense Safari Survey Sheet

Animal number ________	Animal number ________	Animal number ________
☐ claws	☐ claws	☐ claws
☐ horns/antlers	☐ horns/antlers	☐ horns/antlers
☐ teeth	☐ teeth	☐ teeth
☐ quills/spikes	☐ quills/spikes	☐ quills/spikes
☐ hard shells/body coverings	☐ hard shells/body coverings	☐ hard shells/body coverings
☐ tails	☐ tails	☐ tails
Animal number ________	Animal number ________	Animal number ________
☐ claws	☐ claws	☐ claws
☐ horns/antlers	☐ horns/antlers	☐ horns/antlers
☐ teeth	☐ teeth	☐ teeth
☐ quills/spikes	☐ quills/spikes	☐ quills/spikes
☐ hard shells/body coverings	☐ hard shells/body coverings	☐ hard shells/body coverings
☐ tails	☐ tails	☐ tails

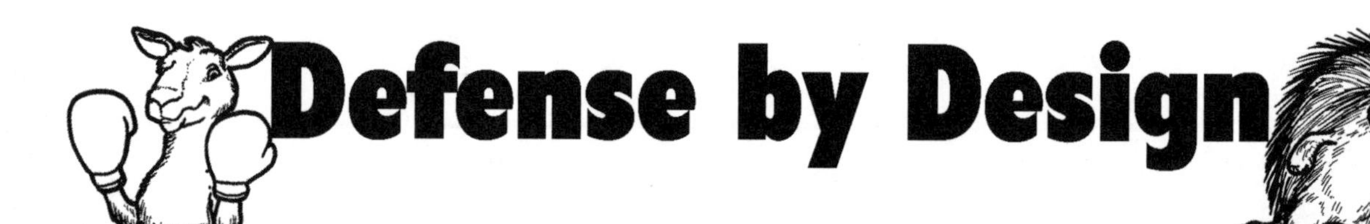

CONNECTING LEARNING

Connecting Learning

1. What are some ways in which animals defend themselves?

2. Why do you think animals have different types of structures for defense?

3. What are some ways claws and teeth can be used other than for defense?

4. How did building the model animal help you learn about animal defenses?

5. What are you wondering now?

The Critter Connection

Camouflage

Camouflage is any use of shape, pattern, and/or color that helps an animal to be less visible. Animals use many different types of camouflage for many different reasons. Some use camouflage to hide from predators. Others use it to hide while hunting prey.

2

There are other kinds of camouflage that animals use, but the four just described are the most common. Find at least one more example of each kind of camouflage and write it below.

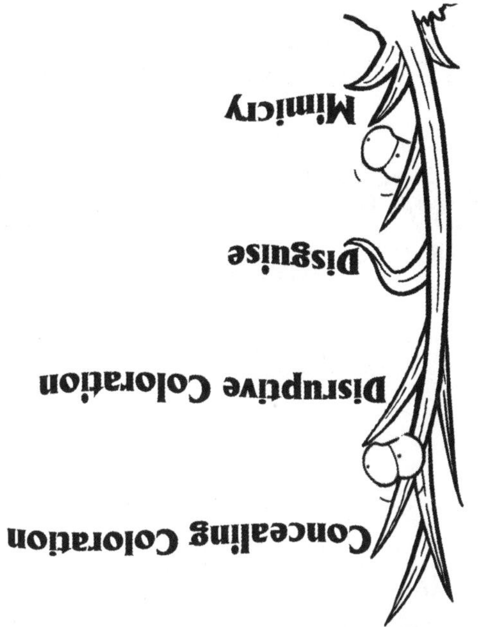

7

Can you think of another type of camouflage that was not defined here? Describe it below.

8

Concealing coloration is a type of camouflage in which animals have body colors that match their surroundings. Animals such as the bobwhite and baby deer have colorings that blend into their normal surroundings. If they change locations, the camouflage no longer works.

Others, like chameleons and octopuses, are able to change the color of their bodies to match their surroundings.

3

Disruptive coloration can be seen in animals like the leopard. Its spots help to disrupt its shape and make it difficult to tell where the leopard ends and its surroundings begin. In order for this kind of camouflage to work, the animal must stay completely still—any movement will give it away.

4

When an animal uses *disguise* as a form of camouflage, it looks like something else. Walking sticks and leaf insects are two examples of animals that are very well disguised just by the shapes of their bodies. There is another insect called an orchid mantis that looks just like one of the flowers on which it sits. It waits for an unsuspecting moth or bee to come to the flowers and then attacks.

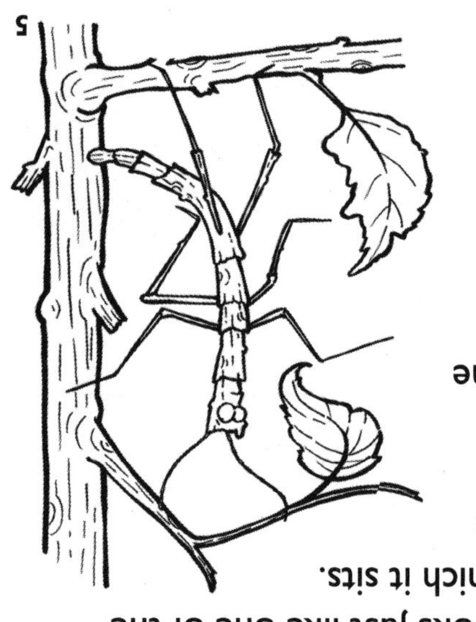

5

Mimicry is another form of camouflage in which one kind of animal looks very much like another animal. The king snake looks very much like the poisonous coral snake that most animals avoid. The viceroy butterfly has markings that are almost identical to the monarch butterfly, which birds don't like to eat because of its bad taste. Both of these animals benefit from looking like something they are not.

6

Topic
Camouflage

Key Question
How does the color of an animal affect its population?

Learning Goal
Students will use paper fish cutouts to see the effect of camouflage on prey populations.

Guiding Documents
Project 2061 Benchmarks
- *Different plants and animals have external features that help them thrive in different kinds of places.*
- *Individuals of the same kind differ in their characteristics, and sometimes the differences give individuals an advantage in surviving and reproducing.*
- *For any particular environment, some kinds of plants and animals survive well, some survive less well, and some cannot survive at all.*
- *Plants and animals have features that help them live in different environments.*

NRC Standard
- *Each plant or animal has different structures that serve different functions in growth, survival, and reproduction. For example, humans have distinct body structures for walking, holding, seeing, and talking.*

*Common Core State Standards for Mathematics**
- *Represent and interpret data. (3.MD)*
- *Develop understandings of fractions as numbers. (3.NF)*

Math
Sequencing
Fractions
Equalities and inequalities
Graphing

Science
Life science
camouflage

Integrated Processes
Observing
Predicting
Comparing and contrasting
Collecting and recording data
Drawing conclusions

Materials
Per group of four:
blue, red, black, and white construction paper
construction paper fish (see *Management 2)*
watch with second hand
student pages

Background Information

For many creatures, color is an important means of defense. The blending of an animal into its environment is called camouflage. Camouflage is one way an animal adapts to its environment. The snowshoe hare is white during the winter so that it can blend in with the snow. When the snow melts, the color of the hare changes to brown so that it blends into the surroundings during the summer months, too. Some animals, like the walking stick and tomato worm, blend so well into their surroundings that they are undetectable when motionless.

Management
1. The fish need to be cut out prior to doing this activity. Students can use the pattern at the top of the second student page or cut out their own fish pattern.
2. The fish need to be the same size and shape and cut from black, red, white, and blue construction paper. Each group will need 12 of each color for the activity.
3. Make a few extra fish in case some get torn during the activity.
4. Small rectangles of paper (2" x 1") can be used instead of the fish shapes.
5. Have groups use two 12" x 18" pieces of blue construction paper for the pond.

Procedure

1. Distribute the student pages.
2. Divide the class into groups of four and make sure each group has the cut-out fish and two sheets of blue paper.
3. Have groups spread the blue paper on a table to act as the fishing pond. Instruct the first "fisherman" to turn his/her back to the pond while the other group members spread the 48 fish evenly over the blue paper.
4. Explain the rules of the activity: Each fishing period will last exactly 10 seconds. The fisherman may only use one hand to pick up fish from the paper. The collected fish may be put in a cup or held in the other hand. The fisherman must attempt to get as many fish as possible in 10 seconds.
5. Have groups time the first fisherman. At the end of 10 seconds, instruct them to count and record the number of fish of each color that were caught.
6. Repeat this procedure until all four people have had a chance to fish. Make sure that all the fish are replaced before starting each new fishing period.
7. Have the groups share their data and make a data chart on the board.
8. Allow time for students to complete the final two student pages. Discuss the results.

Connecting Learning

1. Which color of fish was caught the most often? Why do you think this is?
2. Which color of fish was caught the least often? Why do you think this is?
3. How did your group data compare to the class data?
4. How does this activity relate to animals living in their natural environment?
5. What would happen to snowshoe hares if there was no snow one winter and the ground stayed brown?
6. What are you wondering now?

Extensions

1. Use other colors of fish and backgrounds and repeat the activity.
2. Repeat the activity with the same colors, but different sizes of fish, to see if size makes a difference.

Curriculum Correlation

Literature

Arnosky, Jim. *I See Animals Hiding.* Scholastic, Inc. New York. 2000.

Goodman, Susan. *Claws, Coats, and Camouflage.* Millbrook Press. Brookfield, CT. 2001.

Powzyk, Joyce. *Animal Camouflage: A Closer Look.* Bradbury Press. New York. 1990.

Language Arts

Write a story about a brown mouse that lived in a white environment.

Art

Do a camouflage art lesson in which the students draw pictures with objects hidden in them.

Research

Find other animals that use camouflage.

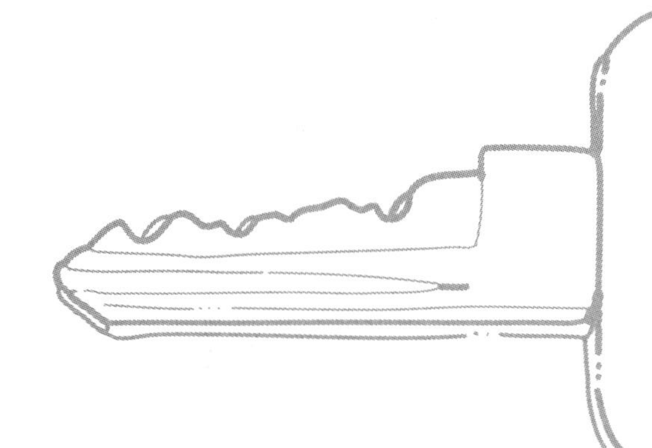

Key Question

How does the color of an animal affect its population?

Learning Goal

Students will:

use paper fish cutouts to see the effect of camouflage on prey populations.

Gone Fishing

Counting our catch

	Group Members 1	2	3	4	Group Totals	Class Totals
White						
Red						
Blue						
Black						

Looking at our catch

1. Which fish was caught the most? __________
   ...the least? __________

2. Label the fish in order from the most caught to the least caught.

Gone Fishing
Cut out this pattern and use it to make your construction paper fish.
Total Number of Fish Caught
White
Red
Blue
Black
Drawing conclusions
Each fish represents ______ fish.

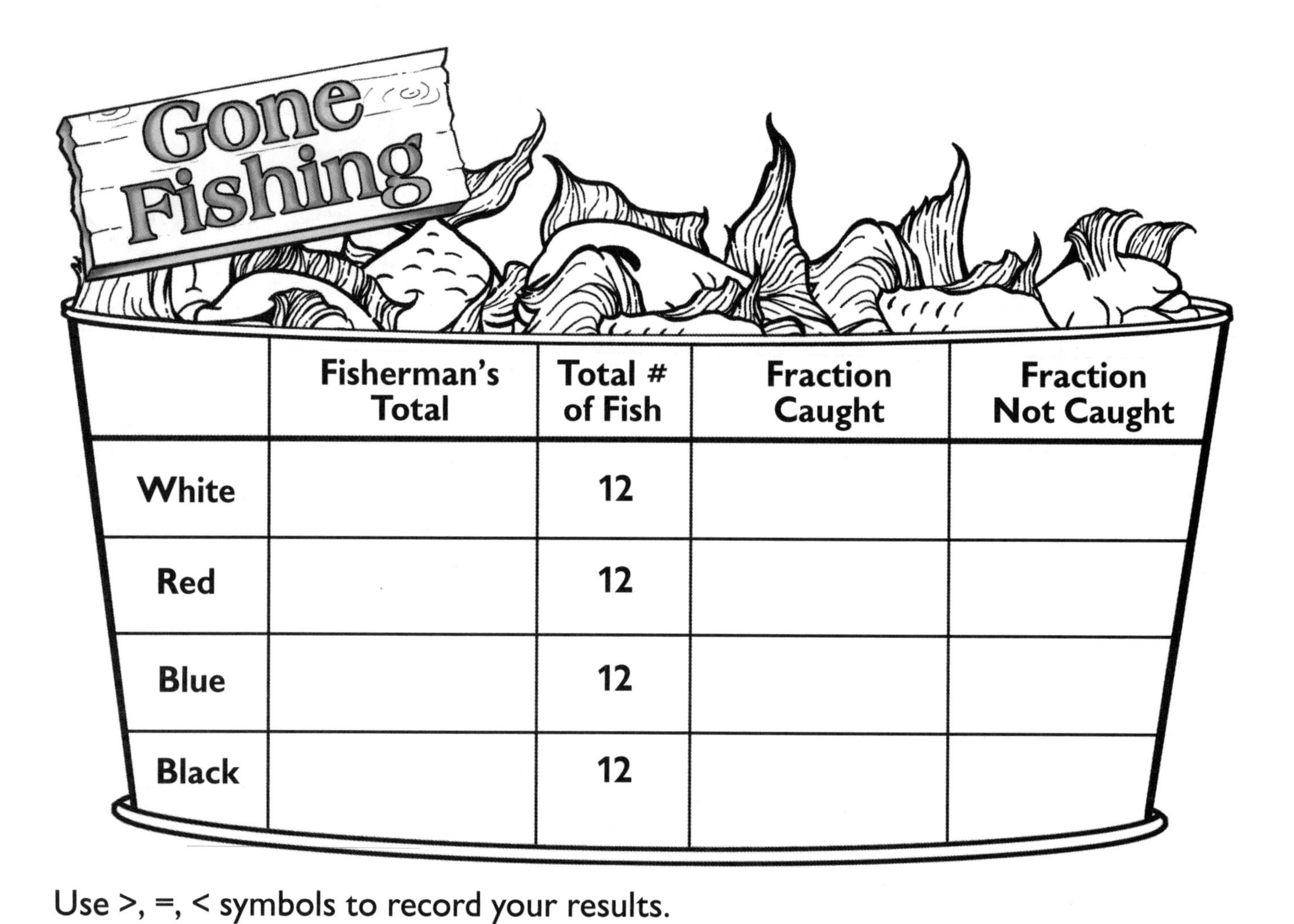

	Fisherman's Total	Total # of Fish	Fraction Caught	Fraction Not Caught
White		12		
Red		12		
Blue		12		
Black		12		

Use >, =, < symbols to record your results.

The fraction of red fish caught is ◯ the fraction of white caught.

The fraction of blue fish caught is ◯ the fraction of black caught.

The fraction of black fish not caught is ◯ the fraction of red fish not caught.

Now fill in the blanks.

The fraction of __________ caught is > the fraction of __________ caught.

The fraction of __________ not caught is < the fraction of __________ not caught.

Write your own sentence about the number of fish caught.

Connecting Learning

1. Which color of fish was caught the most often? Why do you think this is?

2. Which color of fish was caught the least often? Why do you think this is?

3. How did your group data compare to the class data?

4. How does this activity relate to animals living in their natural environment?

5. What would happen to snowshoe hares if there was no snow one winter and the ground stayed brown?

6. What are you wondering now?

Topic
Camouflage

Key Question
How does camouflage affect a critter's ability to survive?

Learning Goals
Students will:
- make a critter, and
- see the effects of camouflage on animal visibility.

Guiding Documents
Project 2061 Benchmark
- *For any particular environment, some kinds of plants and animals survive well, some survive less well, and some cannot survive at all.*

NRC Standard
- *An organism's patterns of behavior are related to the nature of that organism's environment, including the kinds and numbers of other organisms present, the availability of food and resources, and the physical characteristics of the environment. When the environment changes, some plants and animals survive and reproduce, and others die or move to new locations.*

*Common Core State Standards for Mathematics**
- *Solve problems involving measurement and estimation of intervals of time, liquid volume, and masses of objects. (3.MD)*

Math
Measurement
  length
  mass

Science
Life science
  adapations
    camouflage

Integrated Processes
Observing
Predicting

Materials
Camouflage materials (see *Management 3)*
Critter-building materials (see *Management 4)*
Computer with projection system
Internet access (see *Management 5)*
Metric rulers
Balances
Masses

Background Information
Some animals use camouflage to protect themselves from predators. Others use camouflage to better sneak up on prey. Color and body shape are two important variables that enable an animal to blend into its environment. Many animals have adapted to the predominant colors in their surroundings. An animal's body shape and size affect its ability to blend into the environment. The patterns of colors on the body covering also affect the animal's chances of being seen.

Management
1. Be sure to choose an area beforehand where students can hide their critters. This activity woks best if the chosen area is not too large and has definite boundaries.
2. Emphasize that students may not bury the critters.
3. Use tempera paint, markers, or other materials for camouflage.
4. For the critter bodies use paper bags, marshmallows, cardboard egg carton parts, clay, or wads of paper glued or taped together. Toothpicks or sticks can be used for arms and legs.
5. You will need a computer with Internet access and a projection system to show students the camouflage video.
6. Metric rulers (item number 1909), balances (item number 1918), and masses (item number 1923) are available from AIMS.

Procedure
1. Ask the *Key Question* and discuss students' responses. Tell students that you are going to show them a short video on some of the camouflage techniques used by underwater animals.
2. Show students the National Geographic video and/ or some BBC videos (see Internet Connections).
3. Distribute the student pages and explain that students will be creating their own critters to see for themselves the importance of camouflage to survival.

4. Take students outside to the area you selected (see *Management 1)* and define its boundaries. Tell the students to look around and select a spot within this area where they will place the camouflaged critter they will be making.
5. Instruct students to make notes on the first student page that will help them as they create their critter—the colors and shapes present in the environment that they will need to match.
6. Return to the classroom and provide students time to create and camouflage a critter using the materials available. Remind them to keep in mind the information they gathered about the environment in which their critter will be placed.
7. When they are finished, have students measure and find the mass of their critters so that they can record the vital statistics. Encourage them to also complete the critter facts and draw pictures of their critters.
8. Tell the students that they will now test how well their camouflage works. Divide them into two groups and explain that one half will be the "predators" while the other half will be the "prey." They will then have the opportunity to switch roles.
9. Allow the students who are hiding their critters (the "prey") to go outside and place them in the selected locations. Remind students that they may not bury their critters. As the students are hiding their critters, have the "predators" complete the first part of the second student page by making their predictions as to how many critters will be found. (You will need to inform them of the total number of critters being hidden.)
10. When the "prey" students are done hiding their critters, take the entire class outside. Have the students who hid the critters stand around the perimeter of the area where the critters are hidden. Tell them that they cannot say anything to the "predators" while they are hunting.
11. Allow the "predators" into the area and allow them about 60 seconds to search for the hidden critters. Have them pick up each critter they find.
12. At the end of the 60 seconds, instruct the "predators" to stop looking and move to the outside of the area. Allow any students whose critters were not discovered to retrieve them.
13. Return to the classroom and place the critters into two piles—found and not found. Repeat the process of hiding and searching for critters with the roles reversed so that all students have the chance to be the "predators."
14. Make a real graph of all of the critters by dividing them into found and not found categories.
15. Have students complete the questions on the second student page, and discuss what kinds of camouflage were the most effective.

Connecting Learning

1. Were you able to locate all of the critters? Why or why not?
2. Look at the critters in the found and not found groups. What do the critters that were found have in common?
3. What do the critters that were not found have in common?
4. If you were to do this again, what would you change about your critter?
5. Why would a critter's coloring be important to its survival?
6. Can you think of an animal that has protective coloring? Explain.
7. What are you wondering now?

Extensions

1. Emphasize the use of shapes for camouflage. Use green and brown construction paper.
2. Select a different environment or season and adapt your critter.

Internet Connections

Underwater Camo
http://video.nationalgeographic.com/video/player/national-geographic-channel/all-videos/av-2340-3040/ngc-underwater-camo.html
This video (6:37) shows a variety of underwater adaptations and camouflage, including those of cuttlefish, the mimic octopus, and the American tree frog tadpole. Note: begins with advertisement.

BBC Nature
http://www.bbc.co.uk/nature/adaptations/Camouflage#p003x6zx
This page has links to 20 videos on camouflage covering a wide variety of animals including ptarmigans, mountain hares, stick insects, chameleons, and geckos. (Not all videos available in all areas.)

Curriculum Correlation

Arnosky, Jim. *I See Animals Hiding.* Scholastic, Inc. New York. 2000.

Goodman, Susan. *Claws, Coats, and Camouflage.* Millbrook Press. Brookfield, CT. 2001.

Kalman, Bobbie. *Camouflage: Changing to Hide.* Crabtree Publishing Company. New York. 2005.

Weber, Belinda. *Animal Disguises.* Kingfisher. Boston, MA. 2007.

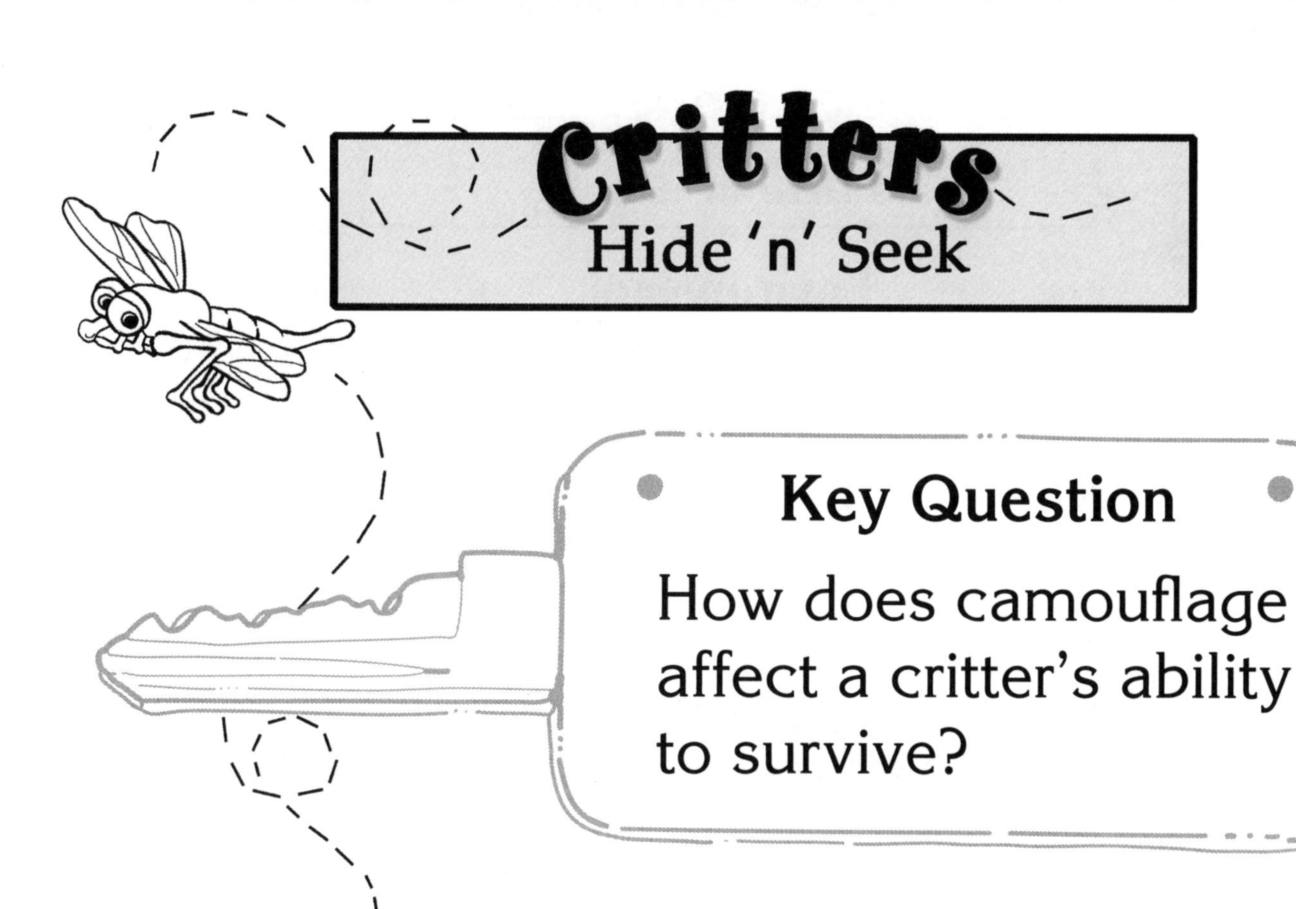

Critters
Hide 'n' Seek

Key Question

How does camouflage affect a critter's ability to survive?

Learning Goals

Students will:

- make a critter, and
- see the effects of camouflage on animal visibility.

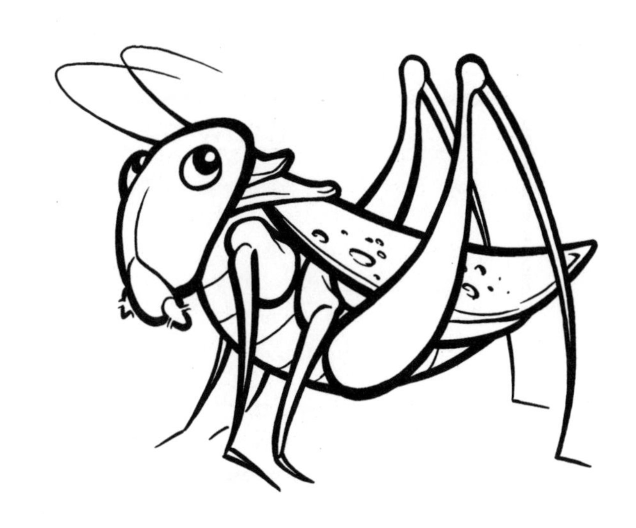

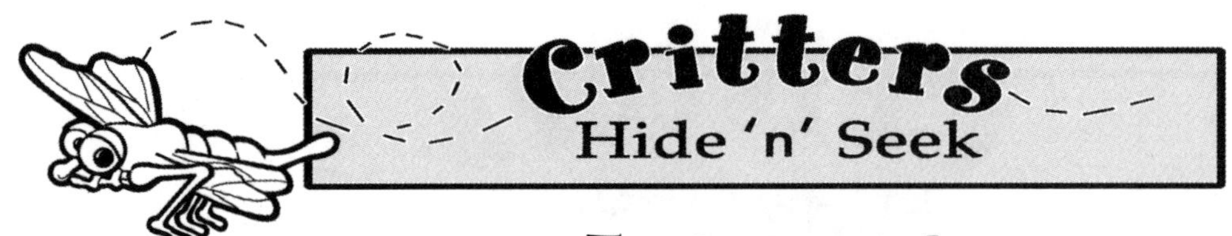

Environment

Observe the environment where your critter will be hiding. Select a specific spot where you want to hide your critter. Record the colors and shapes you see in this spot.

Colors ______________________ ______________________

Shapes ______________________ ______________________

Vital Statistics

Give your critter a name. Measure and record it's height, length, and mass.

Name: ______________________________________

Height __________ **Length** __________ **Mass** __________

Critter Facts

Think about your critter and where it lives. What kind of animals does it eat? What does it eat? What does it use to defend itself? How does it use camouflage?

Predators ______________________

Food ______________________

Defense ______________________

Camouflage ______________________

Draw a picture of your critter here.

Critters Hide 'n' Seek

You will play the part of a predator looking for the critters that are hiding. If you find at least one critter, you will survive. If not, you will starve.

Before you hunt, record the total number of critters and make a guess about how many critters will be found by all the predators.

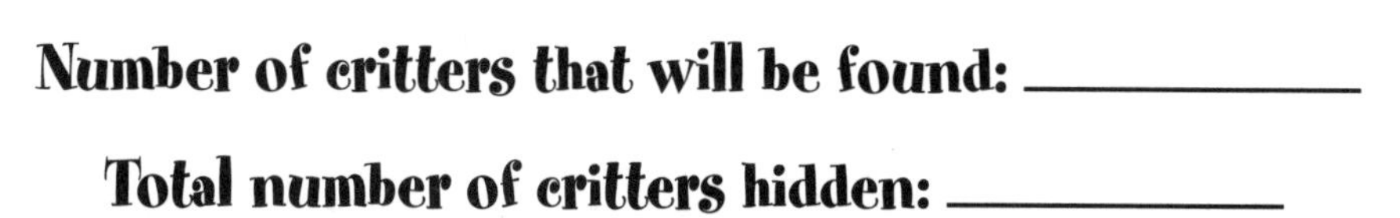

Number of critters that will be found: ____________

Total number of critters hidden: ____________

After the hunt, record the actual results and answer the questions.

Number of critters found: ________ **Number of critters not found:** ________

1. How many critters did you find? Did you survive?

2. Compare the critters that were found and those that were not found. What do the ones that were found have in common?

3. How are the ones that were not found alike?

4. What method of camouflage was most effective? Why?

CONNECTING LEARNING

Connecting Learning

1. Were you able to locate all of the critters? Why or why not?
2. Look at the critters in the found and not found groups. What do the critters that were found have in common?
3. What do the critters that were not found have in common?
4. If you were to do this again, what would you change about your critter?
5. Why would a critter's coloring be important to its survival?
6. Can you think of an animal that has protective coloring? Explain.
7. What are you wondering now?

Topic
Camouflage

Key Question
How does camouflage affect an animal's ability to be seen?

Learning Goals
Students will:
- observe an environment with a variety of moths to see the effects of camouflage on animal visibility,
- design their own camouflaged moths,
- apply this knowledge to the sphinx moth of England and its rapid adaptation during the industrial revolution,
- try to locate moths camouflaged on a coordinate grid, and
- map the moths' locations.

Guiding Documents
Project 2061 Benchmarks
- *For any particular environment, some kinds of plants and animals survive well, some survive less well, and some cannot survive at all.*
- *Different plants and animals have external features that help them thrive in different kinds of places.*
- *Plants and animals have features that help them live in different environments.*
- *Changes in an organism's habitat are sometimes beneficial to it and sometimes harmful.*
- *Individuals of the same kind differ in their characteristics, and sometimes the differences give individuals an advantage in surviving and reproducing.*

NRC Standards
- *Biological evolution accounts for the diversity of species developed through gradual processes over many generations. Species acquire many of their unique characteristics through biological adaptation, which involves the selection of naturally occurring variations in populations. Biological adaptations include changes in structures, behaviors, or physiology that enhance survival and reproductive success in a particular environment.*
- *Each plant or animal has different structures that serve different functions in growth, survival, and reproduction. For example, humans have distinct body structures for walking, holding, seeing, and talking.*

*Common Core State Standards for Mathematics**
- *Model with mathematics. (MP4)*
- *Represent and interpret data. (3.MD)*

Math
Counting
Estimation
Graphing
Computation
 fractions
 decimals
 percents

Science
Life science
 adaptations
  camouflage

Integrated Processes
Observing
Comparing and contrasting
Collecting and recording data
Interpreting data
Drawing conclusions
Applying

Materials
For the class:
 newspaper want ads
 construction paper (brown, green, and white)
 bulletin board paper
 scissors
 tape
 meter stick
 black marker
 computer with projection system
 Internet access (see *Management 3)*

For each student:
 Moths Respond to a Changing Enviroment rubber band book
 rubber bands, #19
 colored pens or pencils
 student pages

Background Information

An animal's ability to blend into an environment is called camouflage. Camouflage can be used for defensive or offensive purposes. A rabbit uses camouflage to hide from predators. A mountain lion uses it to hide until prey are close enough to attack.

Both color and shape can camouflage animals. A walking stick is an example of this. Its shape and color make it appear to be part of a tree branch.

One of the most dramatic cases of an animal's response to a changing environment is the sphinx moth of England. Prior to 1850, the sphinx moths that lived near Manchester were light colored, making them blend into the light-colored bark of the surrounding trees. By 1894, 95% of the sphinx moths were dark colored. This change occurred because of the environmental effects of the industrial revolution. The local industries were burning large amounts of fuel that produced a new phenomenon, air pollution. The vegetation became coated with this pollution and turned darker in color. The light-colored moths became highly visible on the darkened trees and were easy targets for their predators. In 1848 the first black moth was captured, and in 47 years, successive populations of moths had adapted to the darker environment. The darker moths are presently more populous than the lighter ones. However, due to ecological efforts and the use of cleaner fuels, the light-colored moth is beginning to make a comeback. This case is unusual because coloration changes generally occur over a much longer period of time. This example clearly illustrates the impact humans have made on the environment.

Time Line

	Industrial revolution		Environment improves
1800	1850	1900	1950
All moths are light	First dark moth found	Most moths (95%) have become dark	Light moths increasing

Management

1. Before doing the lesson, prepare the "moth environment." To do this, use a double page of the want ads. Be sure the pages are covered with small print. Use the moth pattern to cut out brown, green, white, and newsprint (cut from the want ads) moths. (If you have access to an Ellison machine, you may find it easier to use a die cut to make the moths.) Randomly glue the moths onto the double page of the wants ads. The numbers of each type of moth may vary. Laminate the sheets if desired.
2. You may want to cut the newsprint moths out of another identical page of want ads and glue them in the exact position from which they were cut out, making them very well camouflaged.
3. You will need a computer with Internet access and a projection system to show students the camouflage video(s).
4. Copy one moth pattern on white paper for each student.
5. For *Part Two*, you will need to make a second moth environment or use the environment from *Part One*. Divide this environment into 16 coordinate sections as shown on the student page for *Part Two*. Use a marker and a meter stick to draw the grid lines.

Procedure

Part One

1. Before the students arrive, place the newspaper with the moths glued to it on a wall, bulletin board, or board in the front of the classroom and cover it with a sheet of bulletin board paper.
2. To begin the lesson, tell the class that you have a page of paper moths under the bulletin board paper. Their task will be to look at the paper for 15 seconds and estimate the total number of moths and the number of different types (colors) of moths.
3. Distribute the student page for *Part One*.
4. Remove the bulletin board paper and allow students to observe the paper for 15 seconds. Re-cover it with the bulletin board paper.
5. Have the students complete the first section by recording their estimates of the number of types of moths and the total number of moths they saw. Discuss their estimations.
6. Uncover the paper and count the actual number of types and the total number of moths. Record this data in section two of the student page.
7. Discuss how the predictions and results compared. Because of camouflage, many students may not have seen the newsprint moths. This should be related to how animals depend on protective coloration to help them survive.
8. Complete section three by having the students count and record the number of each type of moth.
9. Complete the bar graph by asking the students to raise their hands when you call out the color of the moth that was easiest for them to see. Record this data on the board and have the students color in their graphs accordingly.
10. Tell them that they are going to choose a spot in the room to place (tape) their moths, but first, they are going to watch some real-life examples of animals that are experts at camouflage.
11. Show one or both of the videos listed in the *Internet Connections*. Discuss how the different fish used camouflage to conceal themselves.
12. Distribute one moth pattern to each student and challenge them to color it so that it will camouflage well in the selected location. Encourage them to consider both color and texture/patterns.

13. Have the students leave the classroom and re-enter one at a time and tape their moths to their chosen locations. Note: Moths must be in plain sight and not be placed under anything.
14. Have all students return and look around to see how many moths they can see.
15. Distribute the rubber band book and read through it together.

Part Two

1. Display the second moth environment at the front of the classroom.
2. Have the students sit so that they are at least two or three meters from the moth environment.
3. Hand out the student pages for *Part Two* and discuss. Guide students as they decide on a key for describing the moths, e.g., Brown = B, Classified = C, etc.
4. Using their keys, have students map the locations of the moths in the moth environment on the student page grid. Do not allow the students to go near the sheet for a closer look. Tell them to mark all of the moths they see on the grid.
5. Point out all of the moths to the class and have them record the actual number of each kind by drawing them into the grid and circling them.
6. Have students write a conclusion about the effectiveness of camouflage.
7. Discuss these conclusions.

Connecting Learning

Part One

1. Which moths were the most camouflaged? ...the least camouflaged?
2. What would have happened if the background had been red? ...black? ...white?
3. How does this example relate to animals in the forest?
4. Why are different moths more easily seen by some of our class?
5. Which of your classmates' moths were you unable to spot? Why? Were others able to spot your moth? Why or why not?
6. Do you think that the sphinx moth changed rapidly? Explain your thinking. [The sphinx moth took between 50 to 100 years to become mostly dark. While on a large scale, this seems rapid, for a moth, this represents many, many generations.]
7. What do you think would have happened to the sphinx moth had it not changed color during the industrial revolution?
8. What are you wondering now?

Part Two

1. How did the moth's color affect its ability to be seen on the grid?
2. Which color was the easiest to locate?
3. Which color was the most difficult to locate?
4. Were there factors other than camouflage that affected your ability to locate each moth?
5. What are you wondering now?

Extensions

1. Use different backgrounds (white, green) for moth environments. Compare results with those using newsprint.
2. Use different colors of moths, or make the moths from another material, such as felt.
3. Have students make camouflage books using old wallpaper patterns. Expired wallpaper books can be obtained from home improvement and home decorating stores.
4. Do this activity on the grass in a marked off area using various shades of green moths.

Internet Connections

Best Disguised Predator Fish
http://video.nationalgeographic.com/video/player/animals/fish-animals/spiny-rayed-fish/stonefish-predation.html
This short video (1:18) shows the camouflaging ability of the stonefish. Note: begins with advertisement.

Sargassum Fish
http://video.nationalgeographic.com/video/player/animals/fish-animals/bony-fish/fish_sargassum.html
This short video (1:42) shows the camouflaging ability of the sargassum fish. Note: begins with advertisement.

Curriculum Correlation

Literature

Arnosky, Jim. *I See Animals Hiding.* Scholastic, Inc. New York. 1995.

Goodman, Susan. *Claws, Coats, and Camouflage.* Millbrook Press. Brookfield, CT. 2001.

Powzyk, Joyce. *Animal Camouflage: A Closer Look.* Bradbury Press. New York. 1990.

Whalley, Paul. *Butterfly & Moth (Eyewitness Books).* DK Publishing, Inc. New York. 2000.

Math

Make a coordinate map of your classroom showing where each desk is located.

Social Sciences

Use a United States map and identify the coordinates of each state's capital.

Key Question

How does camouflage affect an animal's ability to be seen?

Learning Goals

Students will:

- observe an environment with a variety of moths to see the effects of camouflage on animal visibility,
- design their own camouflaged moths,
- apply this knowledge to the sphinx moth of England and its rapid adaptation during the industrial revolution,
- try to locate moths camouflaged on a coordinate grid, and
- map the moths' locations.

Missing Moths
Patterns

Missing Moths

Part One

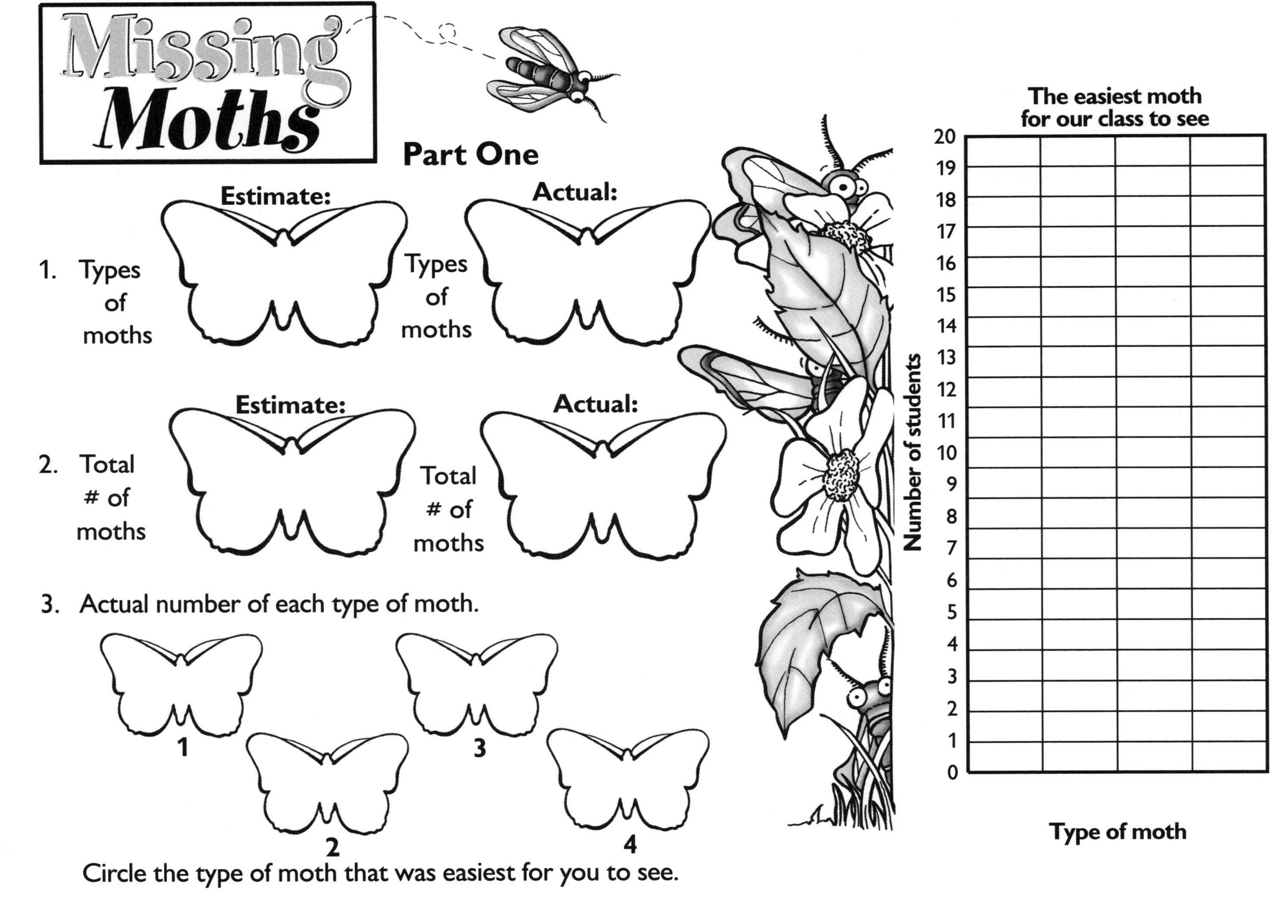

1. Types of moths — Estimate: — Actual: Types of moths

2. Total # of moths — Estimate: — Actual: Total # of moths

3. Actual number of each type of moth.

1 2 3 4

Circle the type of moth that was easiest for you to see.

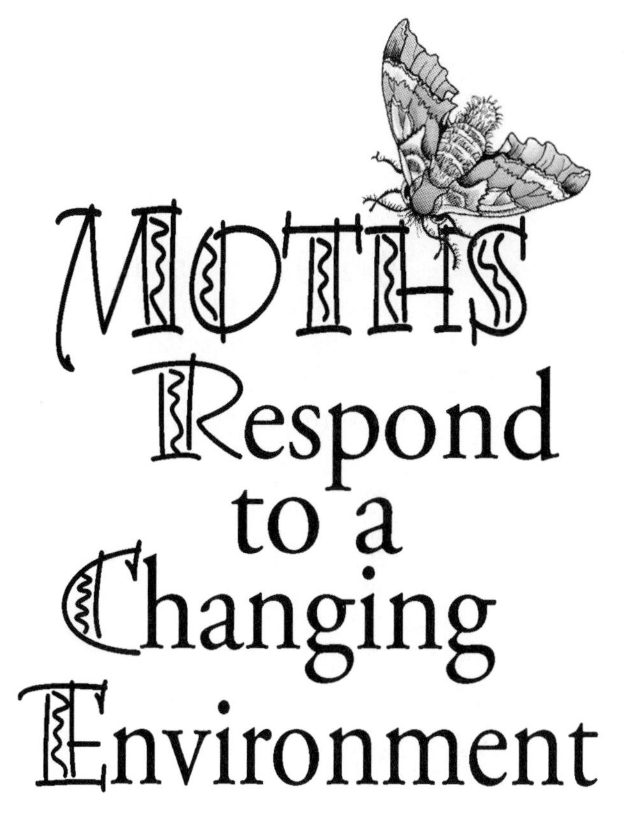

Moths Respond to a Changing Environment

1

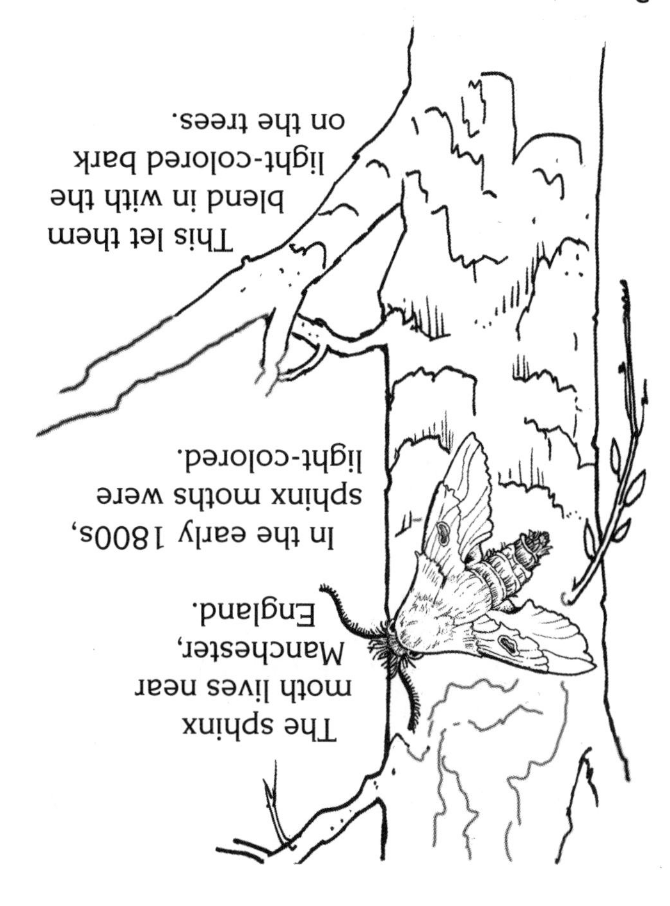

The sphinx moth lives near Manchester, England.

In the early 1800s, sphinx moths were light-colored.

This let them blend in with the light-colored bark on the trees.

2

This case is unusual. Color changes in animals usually take much longer. This example shows the impact humans have made on the environment.

7

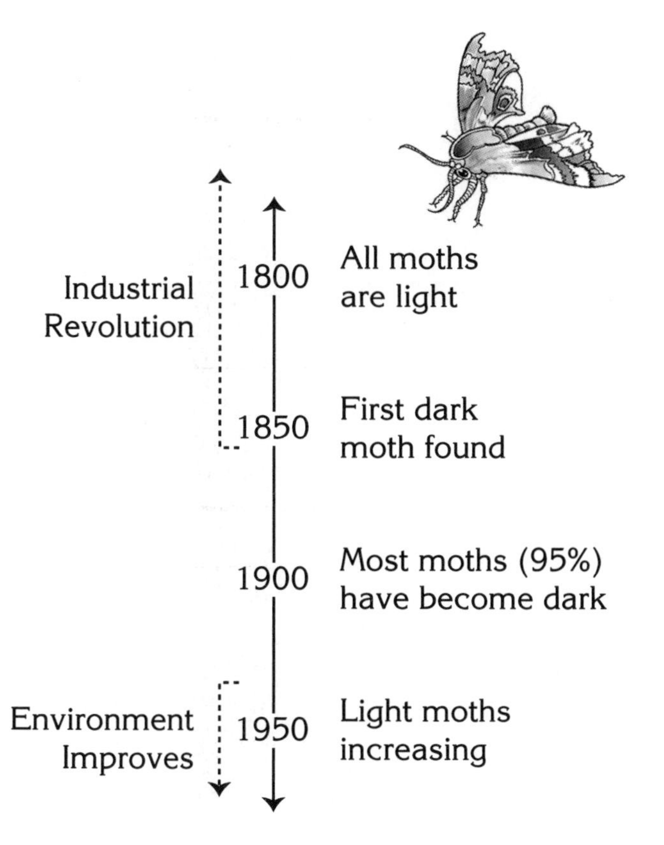

8

By the mid 1800s, changes were happening. Lots of factories were being built. These factories burned coal. The coal smoke turned the bark of the trees dark.

3

4

The light-colored moths became very visible on the dark trees. They did not blend in anymore. They were easy for birds to see.

5

In 1848, the first dark-colored moth was caught. By 1894, 95% of the sphinx moths in Manchester were dark-colored. In just 46 years, the moths changed color. They adapted to the darker trees.

Today there are still more dark-colored moths in Manchester, England than light-colored moths. But now factories use cleaner fuels. The light-colored moths are becoming common again.

6

Using the information from page one, complete the table and graph.

Moth type	Estimate	Actual	Estimate / Actual	Decimal equivalent	Percent estimated

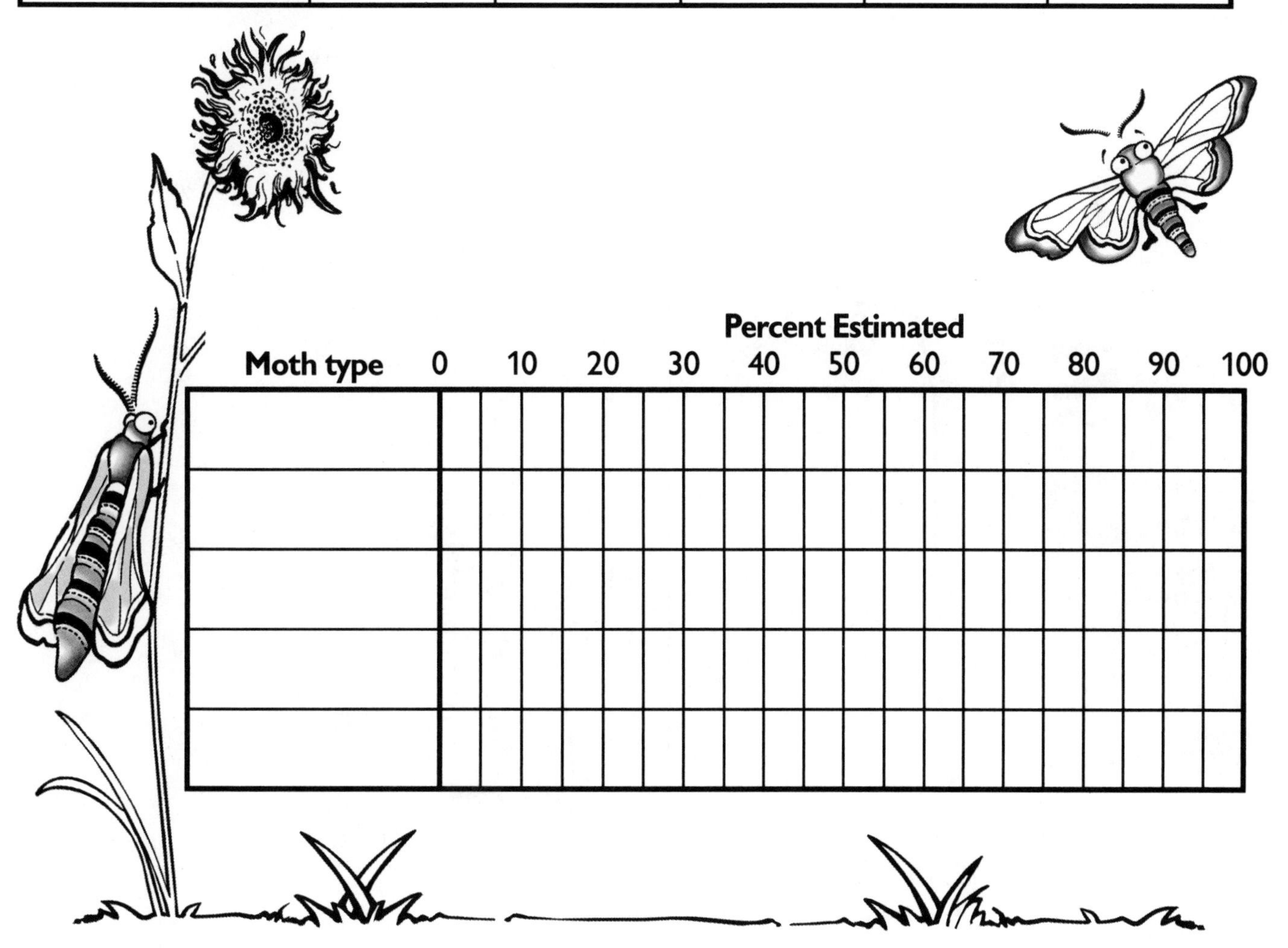

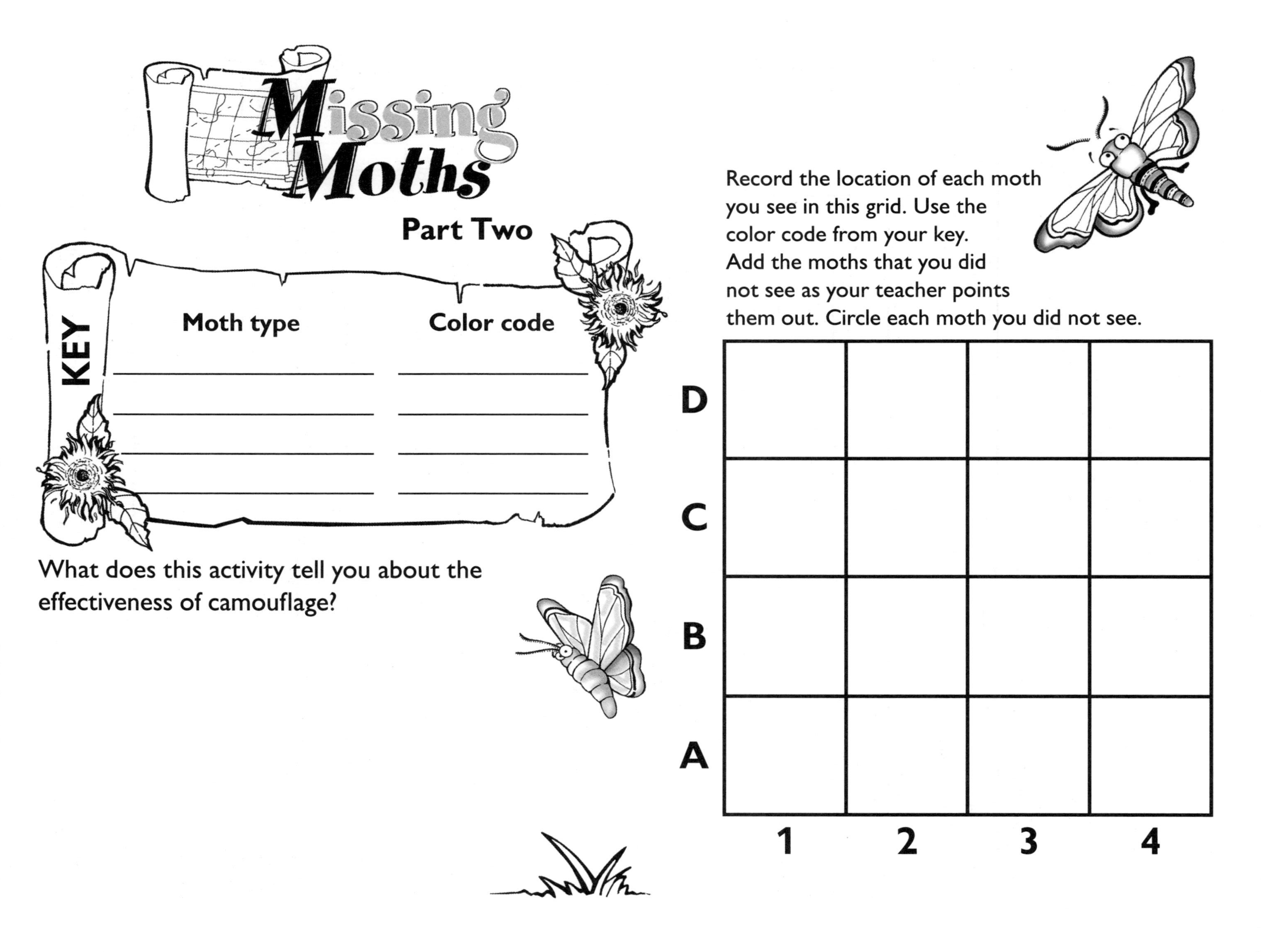

Missing Moths

Part Two

KEY

Moth type	Color code

What does this activity tell you about the effectiveness of camouflage?

Record the location of each moth you see in this grid. Use the color code from your key. Add the moths that you did not see as your teacher points them out. Circle each moth you did not see.

D				
C				
B				
A				
	1	**2**	**3**	**4**

Connecting Learning

Part One

1. Which moths were the most camouflaged? ...the least camouflaged?

2. What would have happened if the background had been red? ...black? ...white?

3. How does this example relate to animals in the forest?

4. Why are different moths more easily seen by some of our class?

5. Which of your classmates' moths were you unable to spot? Why? Were others able to spot your moth? Why or why not?

6. Do you think that the sphinx moth changed rapidly? Explain your thinking.

Connecting Learning

7. What do you think would have happened to the sphinx moth had it not changed color during the industrial revolution?

8. What are you wondering now?

Part Two

1. How did the moth's color affect its ability to be seen on the grid?

2. Which color was the easiest to locate?

3. Which color was the most difficult to locate?

4. Were there factors other than camouflage that affected your ability to locate each moth?

5. What are you wondering now?

"Vore"-acious Eaters

Topic
Needs of organisms

Key Question
What can you learn about the food and water needs of organisms while playing a game?

Learning Goals
Students will:
- identify how the needs of water and food can be met in an environment;
- use data to determine if an animal can survive in an environment; and
- describe the food needs of carnivores, herbivores, and omnivores.

Guiding Documents
Project 2061 Benchmark
- *For any particular environment, some kinds of plants and animals survive well, some survive less well, and some cannot survive at all.*

NRC Standard
- *Organisms have basic needs. For example, animals need air, water, and food; plants require air, water, nutrients, and light. Organisms can survive only in environments, and distinct environments support the life of different types of organisms.*

*Common Core State Standards for Mathematics**
- *Model with mathematics. (MP4)*
- *Represent and interpret data. (3.MD)*

Math
Data analysis
bar graphs

Science
Life science
animals

Integrated Processes
Observing
Communicating
Collecting and recording data
Interpreting data

Materials
For the class:
Animal Needs Cards (see *Management 3)*
one box of multi-colored loop cereal

For each student:
one small piece of clay
six long coffee stirrers
colored pencils or crayons
small lunch bag
student page

Background Information
Organisms' needs must be met in order for them to be able to survive in a specific environment. Organisms need space, food, shelter, water, and air in order to survive in an environment. This activity explores two of life's needs of organisms, water and food.

One way in which animals are grouped is by the food that they eat. Carnivores eat only meat, herbivores eat only plants, and omnivores eat both meat and plants.

Key Vocabulary
Carnivore: meat eater
Herbivore: plant eater
Omnivore: eats both plants and meat

Management
1. Select an area in which to distribute the multi-colored cereal loops that is free from objects that students could trip over.
2. This activity was written for cereal with the following six colors: red, orange, yellow, green, blue, and purple. If the cereal you choose has different colors, you will need to modify the student page.
3. Copy the page of *Animal Needs Cards* enough times so that each student can have one card. Cut apart the cards prior to beginning the activity.

Procedure
1. Ask the *Key Question* and state the *Learning Goals.*
2. Discuss with the students what they ate for their last meal. Point out that they ate the types of food needed for their survival.
3. Inform the students that they will be playing a game that will help them understand that animals must have certain types of food in order to survive in a specific environment.
4. Point out the area that you will use to distribute the multi-colored cereal loops.
5. Tell the students that they will have three minutes to gather the cereal in their lunch bags.

6. Have the students gather the cereal for three minutes. Direct them to return to their seats.
7. Distribute the clay and six coffee stirrers to each student. Tell the students to divide the clay into six equal parts and insert a coffee stirrer into each piece. Have them place the clay balls with the coffee stirrers onto the graphing sheet.
8. Tell the students to place the multi-colored cereal loops onto the coffee stirrers to create a three-dimensional bar graph of what type of loops they collected and record the results on the two-dimensional bar graph.
9. Randomly distribute one *Animal Needs* card to each student.
10. Instruct all students to determine if their food and water needs were met and if they survived.

Connecting Learning

1. What are the two needs that this activity helped you understand that organisms must have met in order to survive? [food, water]
2. What would you have done differently if you could play the game again?
3. This game models what happens when the needs of organisms are not met. Why is it important that all organisms not survive in an environment? [The environment would become overcrowded, and there would not be enough resources to meet their needs.]
4. Which type of eater had the best chance of surviving? [omnivores] Why? [Omnivores eat plants and animals, so they could count more colors as food.]
5. What types of food were represented by the orange, red, green, and yellow loops?
6. What colors were used to represent water?
7. Would there be a problem if there was too much food in an environment? Explain. [maybe if the eaters ate too much and became overweight]
8. What are you wondering now?

"Vore"-acious Eaters

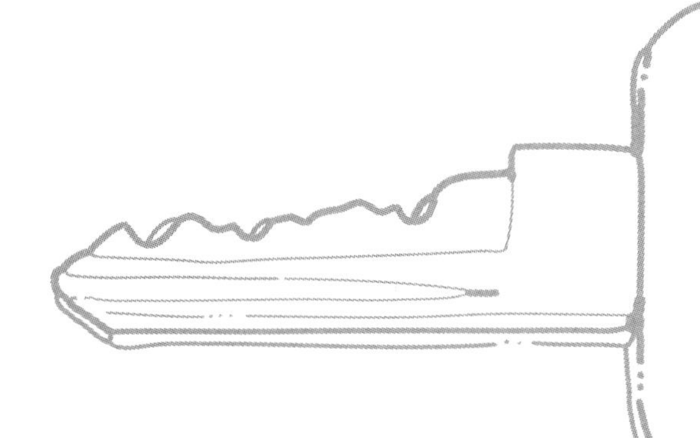

Key Question

What can you learn about the food and water needs of organisms while playing a game?

Learning Goals

Students will:

- identify how the needs of water and food can be met in an environment;
- use data to determine if an animal can survive in an environment; and
- describe the food needs of carnivores, herbivores, and omnivores.

"Vore"-acious Eaters

Animal Needs Cards

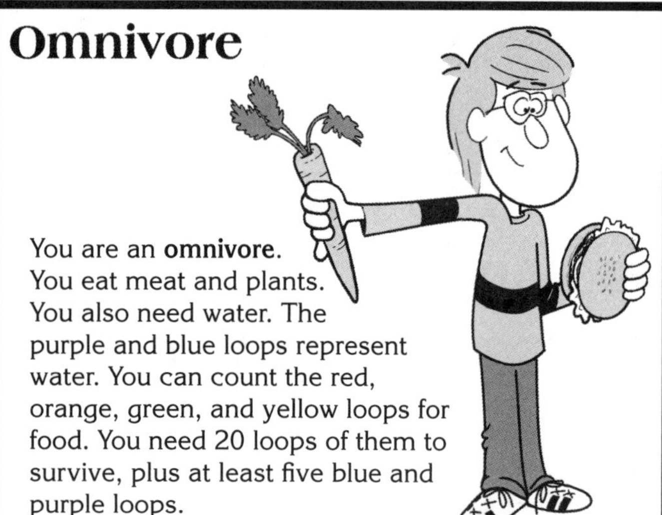

Omnivore

You are an **omnivore**. You eat meat and plants. You also need water. The purple and blue loops represent water. You can count the red, orange, green, and yellow loops for food. You need 20 loops of them to survive, plus at least five blue and purple loops.

Omnivore

You are an **omnivore**. You eat meat and plants. You also need water. The purple and blue loops represent water. You can count the red, orange, green, and yellow loops for food. You need 20 loops of them to survive, plus at least five blue and purple loops.

Herbivore

You are an **herbivore**. You only eat plants. You also need water. The purple and blue loops represent water. The only loops you can count for food are the green- and yellow-colored loops. To survive, you need 20 green and yellow loops and an even number of blue and purple loops.

Herbivore

You are an **herbivore**. You only eat plants. You also need water. The purple and blue loops represent water. The only loops you can count for food are the green- and yellow-colored loops. To survive, you need 20 green and yellow loops and an even number of blue and purple loops.

Carnivore

You are a **carnivore**. You only eat meat. You also need water. The purple and blue loops represent water. The only loops that you can count for food are the red- and orange-colored loops. To survive, you need a total of 15 red and orange loops and an odd number of purple and blue loops.

Carnivore

You are a **carnivore**. You only eat meat. You also need water. The purple and blue loops represent water. The only loops that you can count for food are the red- and orange-colored loops. To survive, you need a total of 15 red and orange loops and an odd number of purple and blue loops.

"Vore"-acious Eaters

Number of Loops

R	O	Y	G	B	P

Loop Color

"Vore"-acious Eaters

Connecting Learning

1. What are the two needs that this activity helped you understand that organisms must have met in order to survive?

2. What would you have done differently if you could play the game again?

3. This game models what happens when the needs of organisms are not met. Why is it important that all organisms not survive in an environment?

"Vore"-acious Eaters

CONNECTING LEARNING

Connecting Learning

4. Which type of eater had the best chance of surviving? Why?

5. What types of food were represented by the orange, red, green, and yellow loops?

6. What colors were used to represent water?

7. Would there be a problem if there was too much food in an environment? Explain.

8. What are you wondering now?

Living things need food to give them energy. A food chain is the path by which energy passes from one living thing to another. All energy originally comes from the sun. Green plants use energy from the sun to make food. Because of this, they are called *producers*.

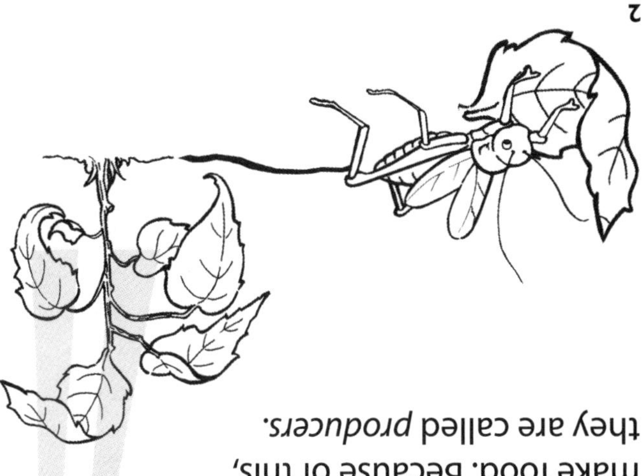

Here is a sample food web. Look at how many food chains are in the food web.

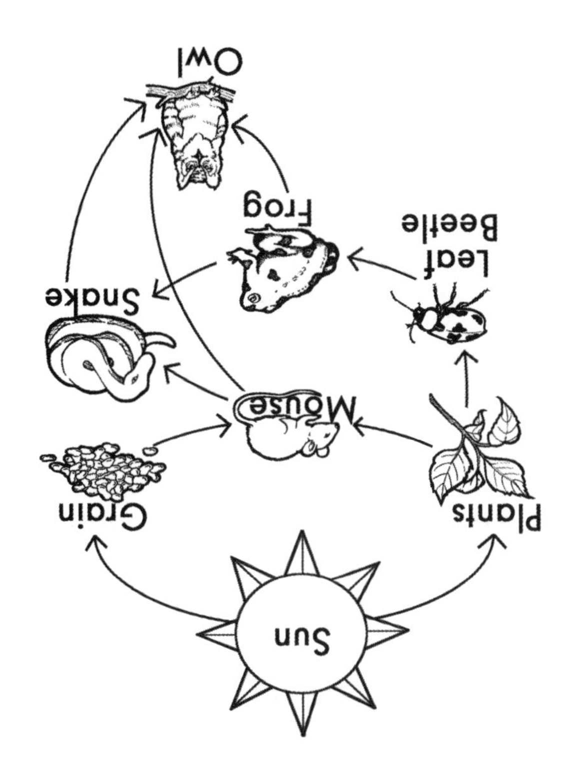

Next time you sit down for a meal, think about where your food is coming from. Are you eating producers, consumers, or both? What kinds of things did the consumers you are eating eat?

Draw a food chain or web that shows something you have eaten recently.

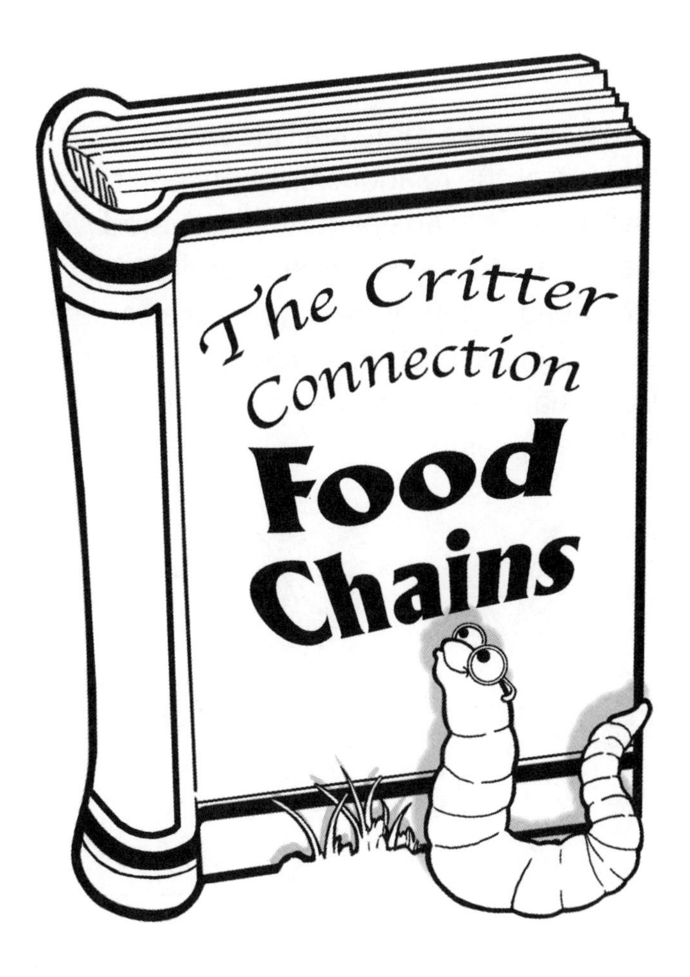

Animals cannot make food using energy from the sun. They must eat plants and/or other animals to get their energy. They are the *consumers* in a food chain. Consumers that eat only plants are called *herbivores.* Those that eat only meat are called *carnivores.* Those that eat both plants and meat are called *omnivores.*

3

Organisms that break down dead or decaying materials into smaller parts are called *decomposers*. Fungi (like mushrooms, mold, and yeast) and bacteria are two kinds of decomposers. Decomposers are an important link in any food chain or web. They return the nutrients that are in a living thing to the soil. Without decomposers, future generations of plants would not have the nutrients they need to grow.

4

A food chain is a simplified way to look at the energy that passes from producers to consumers. One food chain might look something like this:

Sun → Grass → Cow → Human

5

A food web is a more realistic way of looking at the relationships between plants and animals in an environment. A food web is created when several food chains are linked together. Predators eat a variety of prey. It is likely that a predator from one food chain would be linked to the prey of a different chain.

6

Topic
Food chains

Key Question
Where do plants and animals get the energy they need to survive?

Learning Goals
Students will:
- learn about food chains by researching animals and where they get their energy,
- create a food chain using the selected animals, and
- identify the variety of ways transfers of food energy can occur from a source in plants through a series of animals.

Guiding Documents
Project 2061 Benchmarks
- *Some source of "energy" is needed for all organisms to stay alive and grow.*
- *Animals eat plants or other animals for food and may also use plants (or even other animals) for shelter or nesting.*
- *Insects and various other organisms depend on dead plant and animal material for food.*
- *Almost all kinds of animals' food can be traced back to plants.*
- *Two types of organisms may interact with one another in several ways: They may be in a producer/consumer; predator/prey, or parasite/host relationship. Or one organism may scavenge or decompose another. Relationships may be competitive or mutually beneficial. Some species have become so adapted to each other that neither could survive without the other.*
- *All organisms, including the human species, are part of and depend on two main interconnected global food webs. One includes microscopic ocean plants, the animals that feed on them, and finally the animals that feed on those animals. The other web includes land plants, the animals that feed on them, and so forth. The cycles continue indefinitely because organisms decompose after death to return food materials to the environment.*
- *Locate information in reference books, back issues of newspapers and magazines, compact disks, and computer databases.*

NRC Standards
- *All animals depend on plants. Some animals eat plants for food. Other animals eat animals that eat the plants.*
- *Organisms have basic needs. For example, animals need air, water, and food; plants require air, water, nutrients, and light. Organisms can survive only in environments, and distinct environments support the life of different types of organisms.*
- *An organism's patterns of behavior are related to the nature of that organism's environment, including the kinds and numbers of other organisms present, the availability of food and resources, and the physical characteristics of the environment. When the environment changes, some plants and animals survive and reproduce, and others die or move to new locations.*

Science
Life science
  food chains

Integrated Processes
Observing
Comparing and contrasting
Collecting and recording data
Organizing
Drawing conclusions

Materials
Paper plates
Paper clips
Scissors
Glue
Brown paper lunch sacks
Set of animal picture cards (included)
Colored pencils

Background Information

A food chain represents the transfer of energy (originating with the sun) from the producer source to a consumer or a series of consumers. For example, a green plant, a leaf-eating insect, and an insect-eating bird would form a simple food chain.

The food web is divided into two broad categories: the grazing web and the detrital web. The grazing web begins with green plants while the detrital web begins with organic debris. Both webs are made up of individual food chains and represent a series of nutritional levels. Green plants, primary producers of food, belong to the first nutritional level, and plant-eating animals belong to the second level. Predators that feed on the plant-eating animals form the third level, and predators that feed on predators belong to the fourth. As the levels rise, the predators become fewer, larger, and fiercer. Seldom are there more than four or five links or levels in a food chain.

Management

1. Students should work together in small groups on this activity.
2. This activity involves research and will need to take place over a period of a week or more. Students will need access to a variety of research sources such as the Internet, reference books, magazines like *National Geographic* or *Ranger Rick,* or computer databases. References should always be cited when used.

Procedure

1. Introduce the idea of the energy needed to live and grow by discussing the importance of food for our own survival. Extend the discussion to include all animals.
2. Distribute the first two student pages, scissors, colored pencils, paper plates, paper clips, and glue. Tell the students to follow the directions to create a model of two simple food chains.
3. Discuss the energy flow from the sun and through each link.
4. Initiate a "grab-bag research" activity. Use small pictures of animals (included). Place six to eight picture cards in a paper lunch sack and have each group draw out one or two animals for research.
5. Provide groups with the time and materials to find out what kinds of food each animal eats and whether it is eaten by other animals. Make certain students cite their sources of information.
6. Direct the groups to use the information gathered to construct food chains using paper plates as the sun and illustrated and titled paper strips to make links showing the path from the original energy source (the sun), to a food source, to a consumer. An example would be: sun → grain → field mouse → owl. One link in each food chain must include an animal drawn from the grab bag.
7. Have groups present their models to the rest of the class.

Connecting Learning

1. What patterns emerged from your research about food chains?
2. Were you able to discover a food chain that did not originate with a plant source?
3. What kinds of animals eat only plants? [herbivores] ...only animals? [carnivores] ...both plants and animals? [omnivores]
4. Are there any plants that serve as consumers (eat something else, such as animals), or are plants always producers (eaten by something else)? [Some plants are consumers. The pitcher plant is an example. Some of its energy comes from its consumption of insects.]
5. How many different food chains can be constructed using the same consumer?
6. How long can a food chain be? [Food chains are rarely longer than five links.] Give an example of a three-link, four-link, and five-link chain.
7. What are the characteristics of the consumers at the higher levels of the food chain? [fewer, larger, fiercer] How do these compare to consumers at the lower levels?
8. What are you wondering now?

Extension

Construct a giant food web that covers a bulletin board or small wall in your classroom. Write a story that explains how all the parts are interdependent.

Curriculum Correlation

Kalman, Bobbie, and Jacqueline Languille. *What Are Food Chains and Webs?* Crabtree Publishing Company. New York. 1998.

Lauber, Patricia. *Who Eats What?* HarperCollins. New York. 1995.

McKinney, Barbara Shaw. *Pass the Energy, Please!* Dawn Publications. Nevada City, CA. 2000.

Reif, Patricia. *The Magic School Bus Gets Eaten: A Book About Food Chains.* Scholastic, Inc. New York. 1996.

Riley, Peter. *Food Chains.* Franklin Watts. New York. 1999.

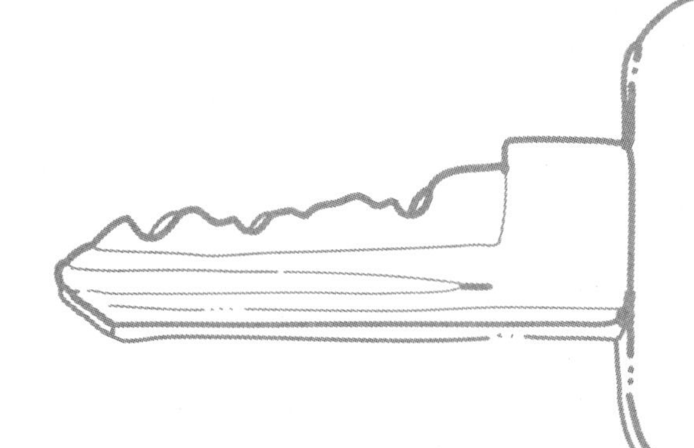

Key Question

Where do plants and animals get the energy they need to survive?

Learning Goals

Students will:

- learn about food chains by researching animals and where they get their energy,
- create a food chain using the selected animals, and
- identify the variety of ways transfers of food energy can occur from a source in plants through a series of animals.

FOOD
Chain
1. Color and cut out the sun. Mount it on a paper plate.
2. Color and cut out the food chains.
3. Glue the links together.
4. Hook the links to the sun with paper clips.

Color the links in the food chains. Cut each strip on the solid lines and glue to make a circle. Make two food chains.

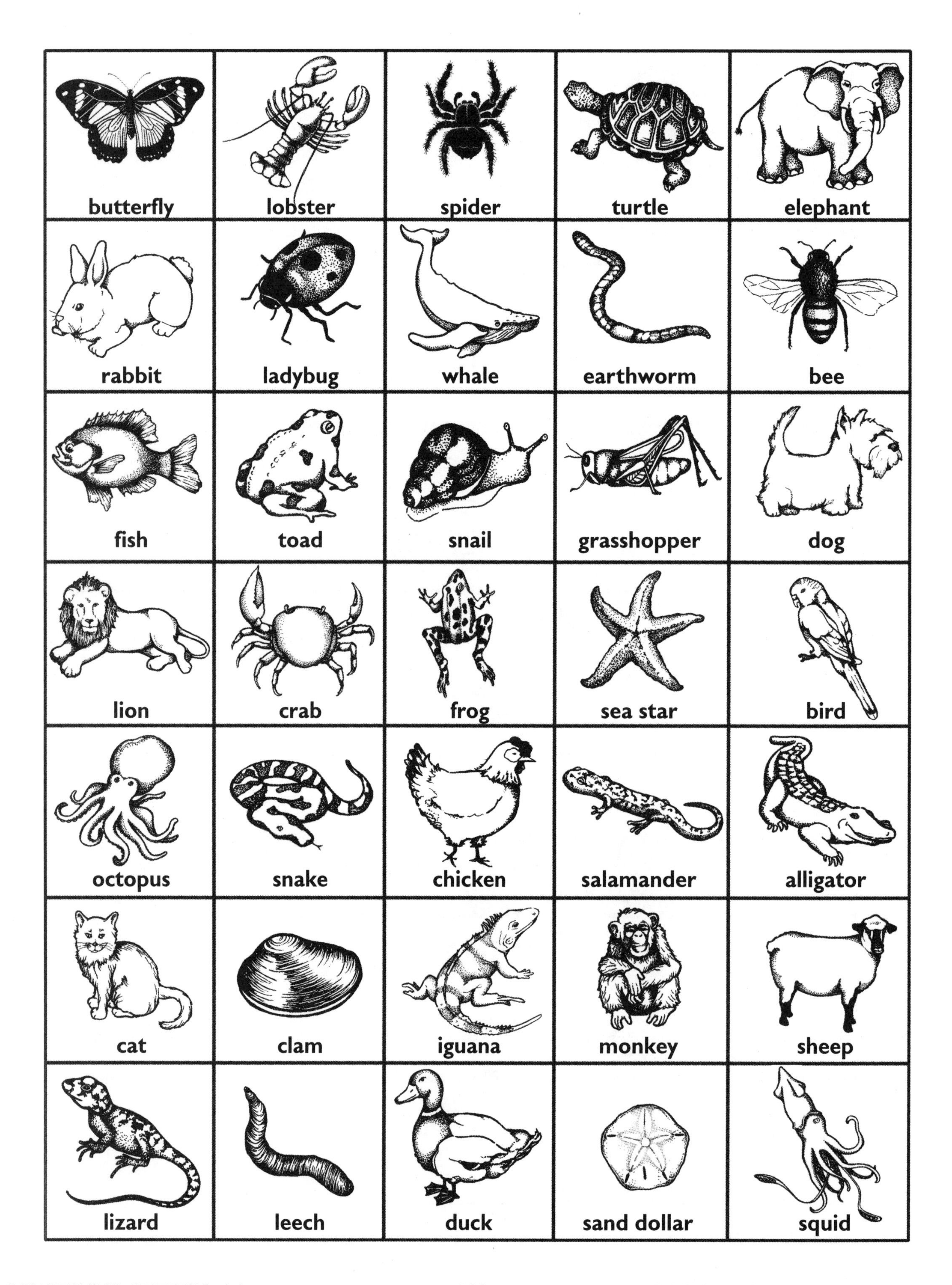
butterfly
lobster
spider
turtle
elephant
rabbit
ladybug
whale
earthworm
bee
fish
toad
snail
grasshopper
dog
lion
crab
frog
sea star
bird
octopus
snake
chicken
salamander
alligator
cat
clam
iguana
monkey
sheep
lizard
leech
duck
sand dollar
squid

Illustrate the links in your food chain. Cut each strip on the solid lines and glue to make a food chain.

glue

glue

glue

glue

glue

CONNECTING LEARNING

Connecting Learning

1. What patterns emerged from your research about food chains?
2. Were you able to discover a food chain that did not originate with a plant source?
3. What kinds of animals eat only plants? ...only animals? ...both plants and animals?
4. Are there any plants that serve as consumers (eat something else, such as animals), or are plants always producers (eaten by something else)?
5. How many different food chains can be constructed using the same consumer?

Connecting Learning

6. How long can a food chain be? Give an example of a three-link, four-link, and five-link chain.
7. What are the characteristics of the consumers at the higher levels of the food chain? How do these compare to consumers at the lower levels?
8. What are you wondering now?

Topic
Food chains

Key Question
What is the primary source of matter and energy entering most food chains?

Learning Goal
Students will know that plants are the main source of matter and energy entering most food chains.

Guiding Documents
Project 2061 Benchmarks
- *Almost all kinds of animals' food can be traced back to plants.*
- *Food provides the fuel and the building material for all organisms. Plants use the energy from light to make sugars from carbon dioxide and water. This food can be used immediately or stored for later use. Organisms that eat plants break down the plant structures to produce the materials and energy they need to survive. Then they are consumed by other organisms.*
- *Some source of "energy" is needed for all organisms to stay alive and grow.*

NRC Standards
- *All animals depend on plants. Some animals eat plants for food. Other animals eat animals that eat the plants.*
- *Populations of organisms can be categorized by the function they serve in an ecosystem. Plants and some microorganisms are producers—they make their own food. All animals, including humans, are consumers, which obtain food by eating other organisms. Decomposers, primarily bacteria and fungi, are consumers that use waste materials and dead organisms for food. Food webs identify the relationships among producers, consumers, and decomposers in an ecosystem.*
- *For ecosystems, the major source of energy is sunlight. Energy entering ecosystems as sunlight is converted by producers into stored chemical energy through photosynthesis. It then passes from organism to organism in food webs.*

Science
Life science
food chains

Integrated Processes
Observing
Comparing and contrasting
Drawing conclusions
Relating

Materials
Box or wastepaper can filled with scratch paper
Labels (see *Management 3)*
Cards
Food Chain Key

Background Information
Food chains exist in all habitats and can be used to demonstrate the energy flow in an ecosystem. Producers capture the sun's energy to make their own food in plant form, while consumers rely on eating those plants or other consumers to get their energy. At each feeding level, there is a 90% loss of energy that was available to the preceding level. Therefore, with each succeeding level having only 10% of the energy available, the number of individuals must decrease. This is why there are so few top-level predators and so many low-level (primary) consumers.

Management
1. This activity is divided into two parts. In the first part, students will play a game in which they toss a ball of wadded scratch paper to see the loss of energy along the food chain. In the second part, students will play a card game in which they build food chains.
2. Copy the cards on card stock. You may wish to laminate the cards for durability.
3. Make labels on paper or 5" x 7" index cards that say *Sun, Grass, Deer, Wolf, Cheese, Mouse, Cat.*
4. The concentration cards can be linked together in several different ways. For example, a hawk might eat a rabbit, a snake, or a smaller bird. Accept any reasonable solutions that the students can explain.
5. Encourage students to form the more common links in the card game. For example, a human could eat a snake and occasionally does, but snakes are not generally a part of a human's diet.

6. The concentration card game can be played by two to four students. Adjust the number of card sets needed by the number of groups you have.

Procedure

Part One

1. Ask the students if they are feeling energetic. Have them name some of the things they have already done today, and ask if doing these things took any energy. See if they can identify a source from which they get all the energy they use. If necessary, guide the discussion to identify food as the energy source, and point out that it takes energy just to do the things we need to stay alive, such as breathing and enabling the other systems of our bodies to function.
2. Ask if they know where the energy in the food comes from, making note of some of the possibilities suggested.
3. Select four students to stand side by side in a line in front of the class. Hand a ball made from four pieces of wadded scratch paper to the first student. Ask the students to pass the ball along the line with each person removing a sheet of the paper. The last person should hold on to the single sheet that is left.
4. Explain that the ball represents energy. Distribute labels to identify the first person as the *sun*, the second as *grass*, the third as a *deer*, and the fourth as a *wolf*. Return the energy ball to the sun and have the students pass it along again as you explain that (1) energy comes from the sun, (2) the grass uses that energy to make and store food in its cells, (3) the deer eats the grass, getting some of the stored energy along with the food matter, and (4) the wolf eats the deer, getting some of the energy that was stored in the deer's cells. Emphasize that the living parts of this arrangement—the plants and the animals—form what is called a *food chain*.
5. Keep the sun and the grass in place, have students return the wadded paper to make a new energy ball, and ask the other two students to return to their seats. Pick three more students, one to represent *cheese* (clarifying that cheese is made from milk that comes from a cow), one to represent a *mouse*, and one to represent a *cat*. Ask the class to arrange the five students in the order of a food chain. When the order is agreed upon, direct the sun to pass the energy ball along the chain again. As the energy ball moves from link to link, have the students remove paper from the energy ball, and ask them to explain how the energy is being passed along from organism to organism.
6. Ask the class to identify what is the same about the two food chains [sun, grass] and what is different [the animals].
7. Tell the students that the entire class will form a circle around the wastepaper can that contains scratch paper. Inform them that the scratch paper represents energy from the sun. Explain that the student who begins the toss will select four or five pieces of scratch paper and wad them up into a ball that he or she tosses to a classmate. When the classmate catches the ball, the classmate must call out what link he or she represents in the food chain. (The first link needs to be a producer such as grass or leaves.) Then that child removes a piece of the paper and tosses the ball to another student. The student who catches it must call out the name of a consumer and remove a piece of the paper. Students will continue tossing and removing the paper until the chain can go no further (usually three to five links). When the chain is finished, tell the students that they will begin again. Invite a student who has not caught the paper ball to go to the wastepaper can and form a new energy ball to begin the process again.
8. After all students have had the opportunity to catch the energy ball and call out a producer or consumer name, inform the class that they will play a card game to look further into food chains.

Part Two

1. Have students shuffle the cards and lay them face down on a desk or table in ordered rows and columns. Distribute a *Food Chain Key* to each group to be used as a reference.
2. This game is played like the commercial "Memory" or "Concentration" games, but instead of looking for matching cards, players are looking for foods chains.
3. Each player's turn begins by turning any two cards face up. If either of those cards begins or continues an existing food chain, the player takes the card(s). For example, any time a player turns over a sun, he/she will take that card because the sun is at the beginning of every food chain.
4. If a player can use both of the cards turned over, he/she continues to turn cards over, one at a time, until a card that he/she cannot use is turned over. At that point, his/her turn is over, the unusable card is placed face down (in the same location) and the next player's turn begins.
5. All food chains must begin with the sun followed by a producer. The subsequent orders of the animals in the food chains can vary, as indicated by the *Food Chain Key*.
6. Players may have multiple food chains going at the same time. They must always keep all food chain cards they have collected face up and arranged in the correct order. This will allow them, and all of the other players, to quickly determine whether or not the card(s) they flip over can be taken.

7. The game ends when there are no more food chains that can be created using the remaining cards. The player with the most cards at the end of the game is the winner.

Connecting Learning

1. Could plants ever be anywhere in a food chain except at the beginning? [No.] Explain. [They can only get their energy directly from the sun, since they make their own food.]
2. Could animals ever be at the beginning of a food chain? Why or why not? [No, because they can't make their own food, so they have to get their energy by eating a plant or another animal.]
3. Which plants have you noticed at the beginning of several different food chains? Why do you think this is so?
4. Which plants are at the beginning of the food chains that include you?
5. What are you wondering now?

Curriculum Correlation

Kalman, Bobbie, and Jacqueline Languille. *What Are Food Chains and Webs?* Crabtree Publishing Company. New York. 1998.
Thorough and informative; explains each role and shows a food web from four different biomes.

Lauber, Patricia. *Who Eats What?* HarperCollins. New York. 1995.
Simple, clear explanation of food chains and webs.

Relf, Patricia. *The Magic School Bus Gets Eaten: A Book About Food Chains.* Scholastic, Inc. New York. 1996.
This story describes an ocean food chain as the connection between pond scum and a tunafish sandwich is explored.

Riley, Peter. *Food Chains.* Franklin Watts. New York. 1998.
Covers the basics about food webs, including focus on five different biomes. Each two page spread has a suggestion for a simple investigation.

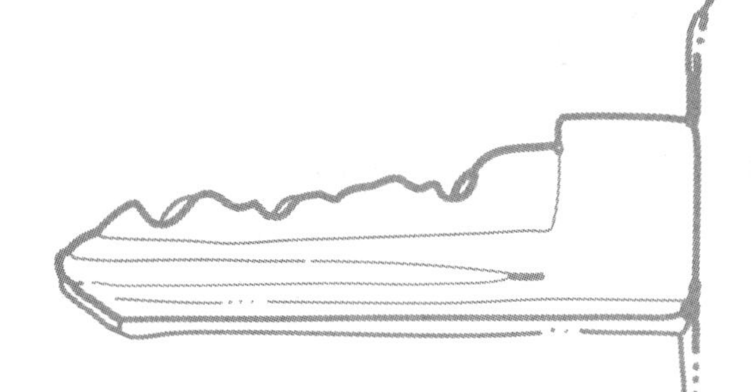

Key Question

What is the primary source of matter and energy entering most food chains?

Learning Goal

know that plants are the main source of matter and energy entering most food chains.

Food Chain Key

Leaf Beetle

What it eats:
Leaves

Cat

What it eats:
Frog, Rabbit, Snake, Mouse, Sparrow

Cow

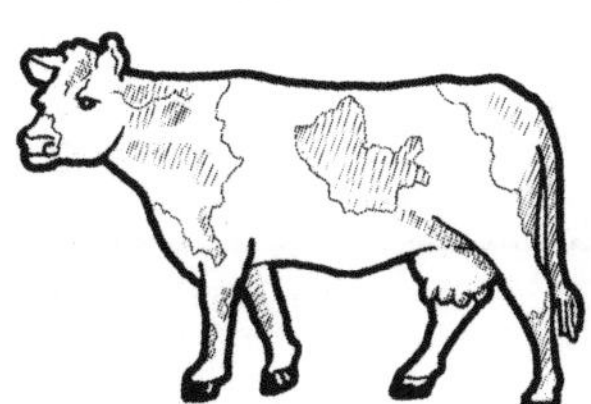

What it eats:
Grain, Grass

Deer

What it eats:
Grain, Grass, Leaves

Frog

What it eats:
Leaf beetle

Hawk

What it eats:
Snake, Frog, Sparrow, Rabbit, Mouse

Man

What it eats:
Cow, Deer, Rabbit, Grain

Mouse

What it eats:
Grain

Owl

What it eats:
Frog, Snake, Sparrow, Mouse, Rabbit

Rabbit

What it eats:
Grain, Leaf, Grass

Snake

What it eats:
Frog, Mouse

Sparrow

Grain, Leaf beetle

CHAIN GAMES

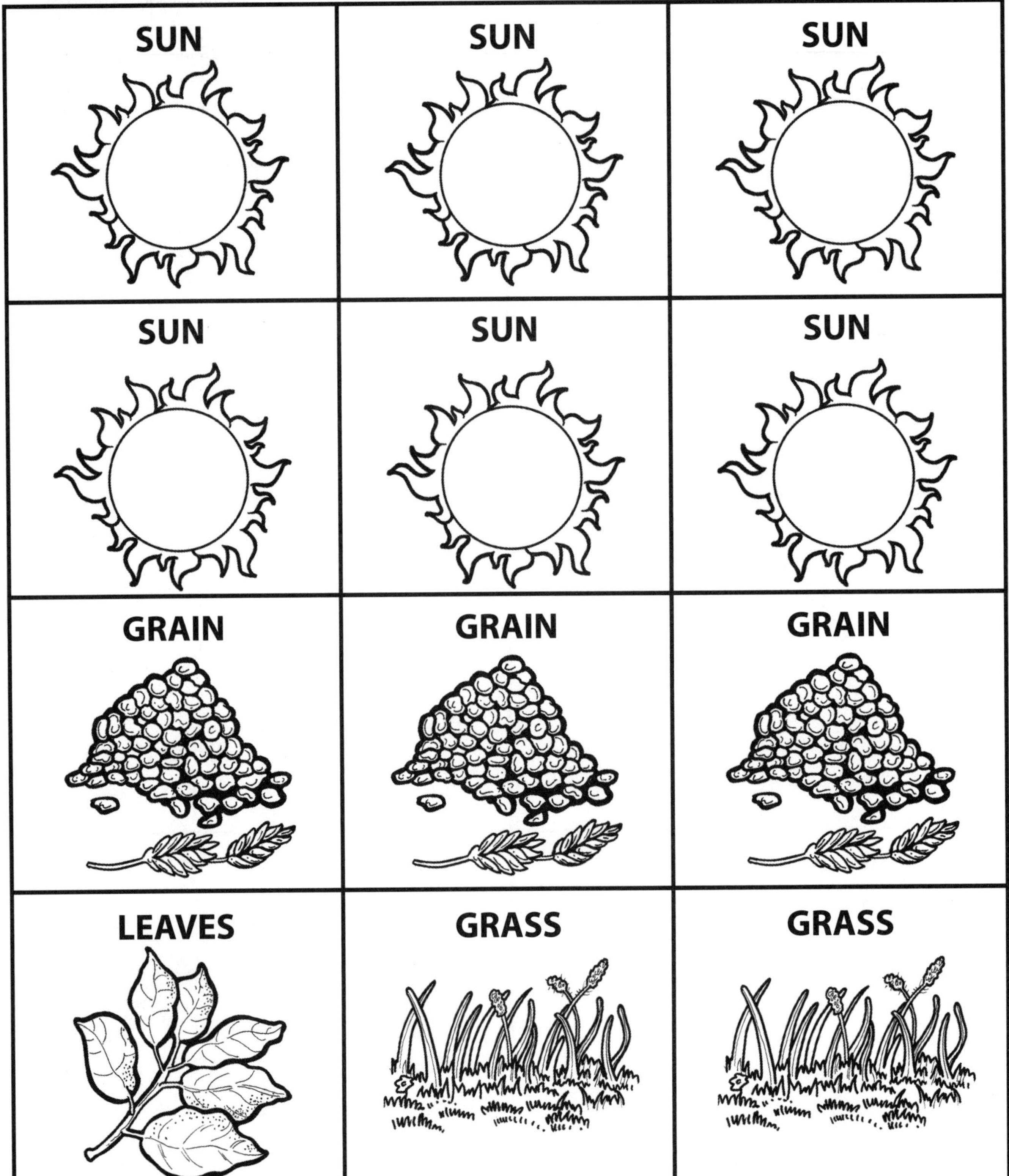
SUN
SUN
SUN
SUN
SUN
SUN
GRAIN
GRAIN
GRAIN
LEAVES
GRASS
GRASS

CHAIN GAMES

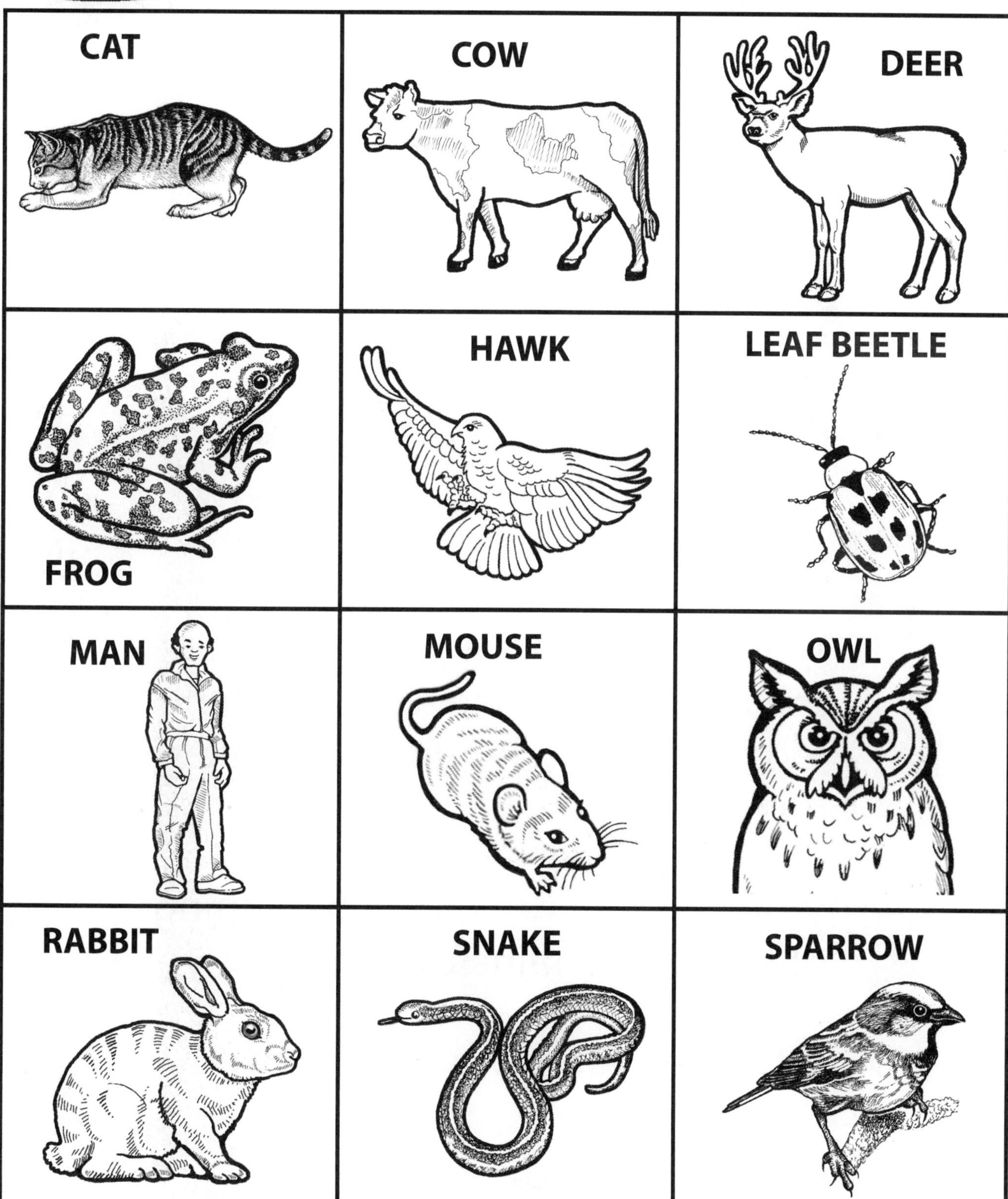
CAT
COW
DEER
FROG
HAWK
LEAF BEETLE
MAN
MOUSE
OWL
RABBIT
SNAKE
SPARROW

CONNECTING LEARNING

Connecting Learning

1. Could plants ever be anywhere in a food chain except at the beginning? Explain.

2. Could animals ever be at the beginning of a food chain? Why or why not?

3. Which plants have you noticed at the beginning of several different food chains? Why do you think this is so?

4. Which plants are at the beginning of the food chains that include you?

5. What are you wondering now?

Topic
Food chains

Key Question
How is energy passed along a food chain from link to link?

Learning Goals
Students will:
- learn about the predator/prey relationship in a food chain, and
- play a game of tag to experience this relationship.

Guiding Documents
Project 2061 Benchmarks
- *Animals eat plants or other animals for food and may also use plants (or even other animals) for shelter or nesting.*
- *Almost all kinds of animals' food can be traced back to plants.*
- *Some source of "energy" is needed for all organisms to stay alive and grow.*
- *Two types of organisms may interact with one another in several ways: They may be in a producer/consumer; predator/prey, or parasite/host relationship. Or one organism may scavenge or decompose another. Relationships may be competitive or mutually beneficial. Some species have become so adapted to each other that neither could survive without the other.*
- *All organisms, including the human species, are part of and depend on two main interconnected global food webs. One includes microscopic ocean plants, the animals that feed on them, and finally the animals that feed on those animals. The other web includes land plants, the animals that feed on them, and so forth. The cycles continue indefinitely because organisms decompose after death to return food materials to the environment.*

NRC Standards
- *All animals depend on plants. Some animals eat plants for food. Other animals eat animals that eat the plants.*
- *Organisms have basic needs. For example, animals need air, water, and food; plants require air, water, nutrients, and light. Organisms can survive only in environments, and distinct environments support the life of different types of organisms.*
- *An organism's patterns of behavior are related to the nature of that organism's environment, including the kinds and numbers of other organisms present, the availability of food and resources, and the physical characteristics of the environment. When the environment changes, some plants and animals survive and reproduce, and others die or move to new locations.*

Science
Life science
  food chains
    predator/prey relationship

Integrated Processes
Observing
Comparing and contrasting
Collecting and recording data
Identifying and manipulating variables
Generalizing
Analyzing

Materials
Brown, yellow, and red yarn (see *Management 3)*
Large bag of plain popped popcorn
Plastic sandwich bags, one per student
Scissors
Thread
Tape
Student page
Cheese popcorn, optional

Background Information
All foods contain chemical energy. A food chain shows the transfer of energy through the chain. Energy is released from the sun and converted by green plants (producers in the food chain) that use light to make food through photosynthesis. Primary consumers are dependent on green plants, and thus the sun, for food energy. Higher level consumers are dependent on the animals that eat plants or other animals, thus the energy is passed from link to link in the food chain. All links are ultimately dependent on the sun for their food energy. Some energy is lost between each link in a food chain. Because of the energy loss, each higher level has fewer living things than the level below it. This means that food chains rarely exceed four or five links.

A pyramid of energy, or biomass pyramid, illustrates the energy transfer between predators and prey. Animals at the top of the pyramid are fewer in number and need to eat many smaller animals to get enough energy to survive. The primary consumers that feed on green plants are much more numerous. In a well-balanced ecosystem, the producers and consumers at each level have numbers that are large enough to ensure their survival without depleting their food supply, thus the pyramid effect with many producers and primary consumers and few of the highest level consumers.

Management

1. Find an area with well-defined boundaries for this outdoor activity.
2. Stress safety and demonstrate the proper way to tag. Make sure students understand the rules before going outside.
3. You will need yarn in brown, red, and yellow cut into pieces about 40 cm long. Cut enough brown for half the class, enough yellow for one-third of the class, and enough red for one-third of the class.
4. Use the student page after the final round of play.

Procedure

1. Review food chains and food webs, if appropriate. Discuss predator/prey relationships.
2. Tell students they are going to play a game of tag that will simulate a natural food chain and illustrate a biomass pyramid.
3. Divide the class into three even groups. Each group will be assigned a different color of yarn. Hand out the yarn and have students help each other tie the yarn around their wrists in a bow that can be easily removed at the end of the game.
4. Explain that the animals the students are simulating are represented by the colors of yarn. Students with brown yarn are grasshoppers, those with yellow yarn are lizards, and those with red yarn are hawks.
5. Discuss the predator/prey relationships in this food chain. Hawks hunt only lizards. Lizards hunt only grasshoppers. Grasshoppers eat only grass (which is represented by the popcorn).
6. Give each student a plastic bag to be used as a stomach and explain how the game will work. Those students who are grasshoppers must gather popcorn from the ground and put in their plastic bags. The students playing lizards will try to tag the grasshoppers. If they are successful, the grasshopper is "dead" and the contents of his/her bag are emptied into the lizard's bag. (The empty bag stays with the grasshopper to be used again in the next round.) The students playing hawks will try to tag the lizards, and get the contents of their bags if successful. Lizards and hawks may not pick up popcorn from the ground.
7. For the animals to survive, they must not be tagged during the game and their stomachs (plastic bags) must be filled by the game's end as follows:
   grasshoppers—1/3 full
   lizards—2/3 full
   hawks—full
8. Go outdoors and select an area to be the ecosystem. For the first round, the area should be small so that the students can experience the effects of crowding on animal populations. Students may not leave the area during the game.
9. Set up two or three safe zones within the selected area. These zones should be large enough for two students at a time. Whenever a new student enters a safe zone, the one who has been there longer must leave. Animals may not prey on each other in these zones. Select another area in which the "dead" animals can wait for the next round.
10. Spread out a large bag of popped popcorn over the ecosystem.
11. Signal the primary consumers, the grasshoppers, to begin eating grass (gathering popcorn). After 30 seconds, allow the lizards to enter the area. After 30 more seconds, allow the hawks to enter the ecosystem. Allow the students to play for several minutes or until no prey are left. At the end of play, all remaining animals must have the right amount of food in their plastic bags or they too are dead. Note the length of time the game lasted.
12. After this first round, ask why the game only lasted a few minutes. Discuss crowding and the number of predators vs. the number of prey. Write down the number and kinds of animals that are still alive.
13. Play the second round in the same area as the first one with the following changes: have half the students play grasshoppers, and divide the other half so that two-thirds of them are lizards and one-third are hawks. Play the game again. Discuss the effects that changing the population numbers had on the time the game lasted.
14. For the third round, leave the animal populations as they were in round two, but greatly enlarge the area in which the game is played. Discuss the effects of the larger area on the time the game lasted.
15. Return to the classroom. Use the activity sheet to illustrate the numbers of predators and prey in an ecosystem and to make a biomass mobile. Discuss the energy flow from the producers to the higher level consumers. Emphasize that the energy in a food chain originates from the sun. A biomass pyramid could also be made by centering and gluing the pieces, one on top of another.

Connecting Learning

1. Why did the games end? [All the prey were dead.] Which games were the shortest? [those with the fewest prey] ...longest? [those with the most prey] Why? [The more prey there are the longer it takes for the predators to eat them all.]
2. What numbers of predators and prey worked the best? [more prey, fewer predators]
3. How does area affect predator/prey relationships?
4. How is this game like a real ecosystem? [It shows one food chain and how the predator/prey relationship works.] How is it different? [There would be more variables in a real ecosystem.]
5. Where does grass get energy? [the sun]
6. Where does the grasshopper get energy? [grass] ...the lizard [grasshopper] ...the hawk? [lizard]
7. How is the mobile related to the predator/prey game? Why are green plants so important in a food chain?
8. What are you wondering now?

Extensions

1. Mix some cheese popcorn in with the regular popcorn to represent a pesticide. Do not tell students what it represents until the end of the game. When the round is over, inform them that any animal with three or more pieces of cheese popcorn in its stomach is dead due to toxic poisoning.
2. Create a variety of food chains using other animals.
3. Play the game introducing predator/prey behaviors such as camouflage, hunting techniques, decoying, running speed, freezing, and playing dead.

Curriculum Correlation

Geography

Research various geographical areas and list several food chains found there that are different than those found in your area.

Art

Design a poster illustrating food chains and food webs.

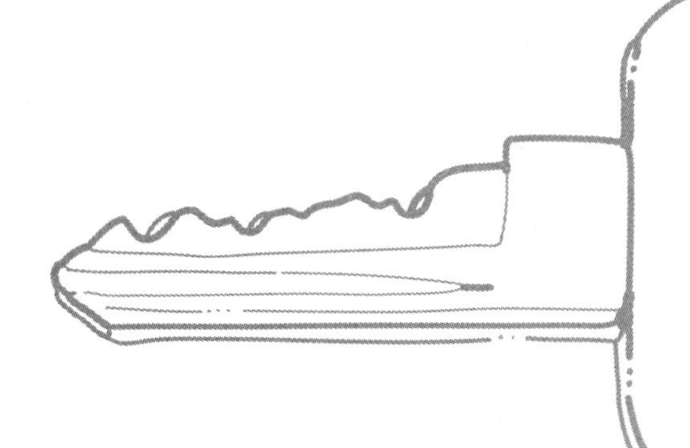

Key Question

How is energy passed along a food chain from link to link?

Learning Goals

Students will:

- learn about the predator/prey relationship in a food chain, and
- play a game of tag to experience this relationship.

Square	Length	Width	Area
A		x	=
B		x	=
C		x	=
D		x	=

Find the length, width, and area of each square below. Record this information in the table.

Imagine that the squares are part of an ecosystem that includes grass, grasshoppers, lizards, and hawks. Think about the numbers of living things in a balanced ecosystem and color the squares according to this key.

lizards = yellow	grasshoppers = brown	grass = green	hawks = red

A

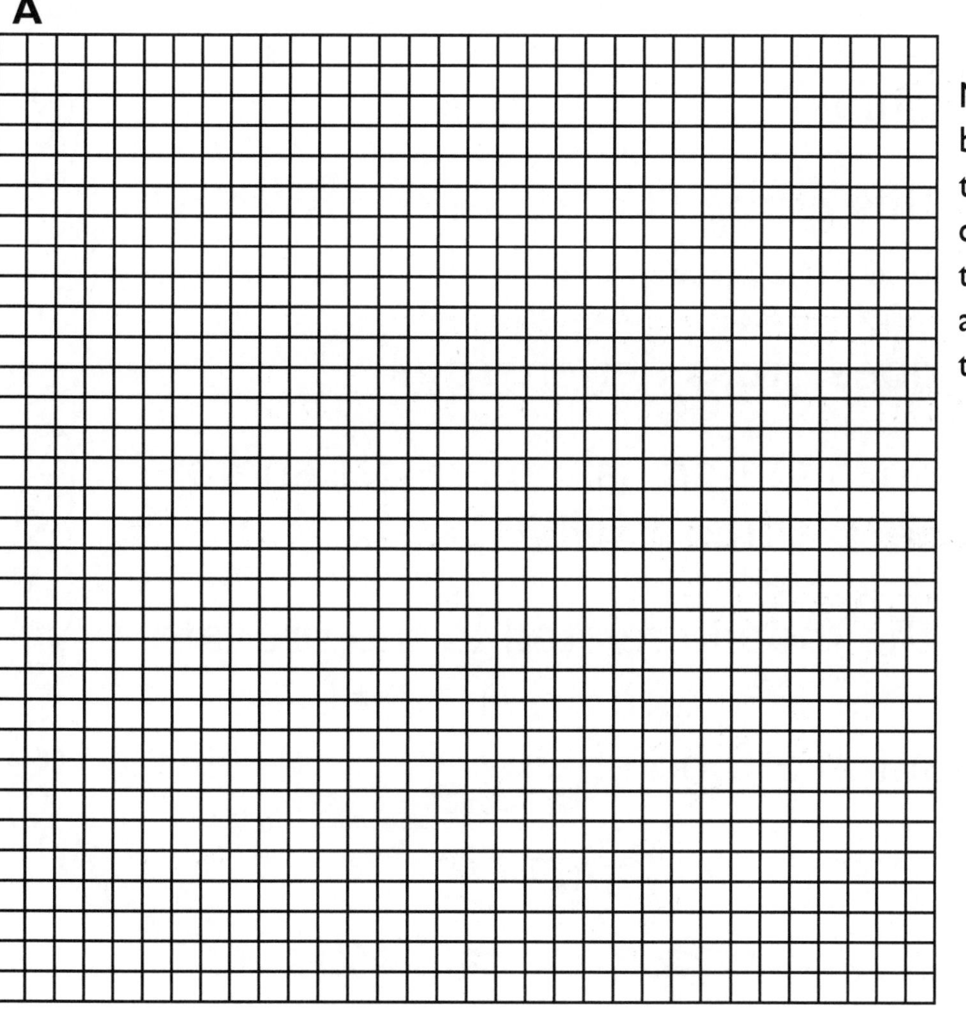

Make a mobile by cutting out the squares and connecting them with a piece of thread.

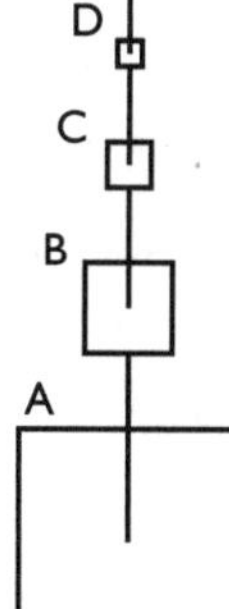

B

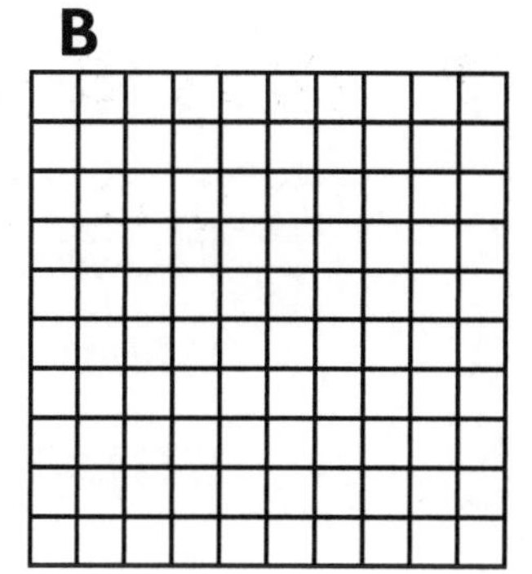

C

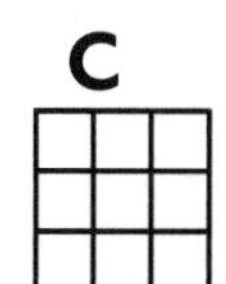

D

CONNECTING LEARNING

Connecting Learning

1. Why did the games end? Which games were the shortest? ...longest? Why?
2. What numbers of predators and prey worked the best?
3. How does area affect predator/prey relationships?
4. How is this game like a real ecosystem? How is it different?
5. Where does grass get energy?
6. Where does the grasshopper get energy? ...the lizard ...the hawk?
7. How is the mobile related to the predator/prey game? Why are green plants so important in a food chain?
8. What are you wondering now?

Topic
Food chains

Key Question
How does a change in the food chain affect the ecosystem?

Learning Goal
Students will learn about how a change in the food chain affects an ecosystem by listening to and/or participating in a reader's theater.

Guiding Documents
Project 2061 Benchmarks
- *Organisms interact with one another in various ways besides providing food. Many plants depend on animals for carrying their pollen to other plants or for dispersing their seeds.*
- *Changes in an organism's habitat are sometimes beneficial to it and sometimes harmful.*
- *Almost all kinds of animals' food can be traced back to plants.*
- *Some source of "energy" is needed for all organisms to stay alive and grow.*

NRC Standards
- *Organisms have basic needs. For example, animals need air, water, and food; plants require air, water, nutrients, and light. Organisms can survive only in environments in which their needs can be met. The world has many different environments, and distinct environments support the life of different types of organisms.*
- *All animals depend on plants. Some animals eat plants for food. Other animals eat animals that eat the plants.*
- *An organism's patterns of behavior are related to the nature of that organism's environment, including the kinds and numbers of other organisms present, the availability of food and resources, and the physical characteristics of the environment. When the environment changes, some plants and animals survive and reproduce, and others die or move to new locations.*

Science
Life science
  food chains
  ecosystems

Integrated Processes
Observing
Relating
Interpreting

Materials
Character necklaces (see *Management 1)*
Reader's theater scripts
What if There Were No Bees? by Suzanne Slade
Music stands, optional

Background Information
Living things are connected by food chains. These chains show where living things get their energy. Green plants make their own food using energy from the sun. Primary consumers eat the green plants, gaining some of their energy. Secondary consumers eat the animals that eat the green plants, and so on. When one part of the food chain is altered, it sets off a chain reaction that affects not only the food chain, but also the entire ecosystem.

In this reader's theater, students will learn about some of the food chains in a grassland ecosystem. They will recognize the importance of a small and seemingly insignificant member of these food chains—the bee. Bees are a keystone species. A keystone species is one on which other species in an ecosystem largely depend. The removal of a keystone species would drastically alter an ecosystem, as students will learn.

Management
1. Copy the character pages and tape a length of yarn to each so that the students can wear them around their necks.
2. You may wish to provide students with music stands on which to place their scripts so that their hands are free to gesture or perform other actions as needed.

3. Select nine students to perform the reader's theater. If you have time, multiple groups can be given the opportunity to perform. Students should be given time to read the script and rehearse together before presenting to the class. This may need to happen over a period of several days.

Procedure

1. Distribute the scripts and character necklaces to the selected students. Allow them time to read the scripts and rehearse together.
2. When students are ready to perform, have the readers come to the front of the class while you set the scene. Tell students that they are going to be watching a reader's theater. It's kind of like a play, but the readers don't memorize the lines or act things out in the same way.
3. Explain that this reader's theater is about the food chains in a grassland ecosystem. Invite students to listen quietly as their classmates perform.
4. If desired, repeat with other groups of students who have also prepared.
5. Read the book *What if There Were No Bees?* by Suzanne Slade.
6. Discuss what students learned about the food chains of a grassland ecosystem and how changes in a food chain affect an ecosystem.

Connecting Learning

1. What is one food chain in the grasslands ecosystem that you learned about in the reader's theater? [fruit → mouse → snake → hawk]
2. Where do the bees fit in this food chain? [They pollinate the plants that make the fruit the mice eat.]
3. What other food chains did you learn about from the book?
4. What would happen to the grassland ecosystem if there were no more bees?
5. What did you like most about this reader's theater?
6. What are you wondering now?

Curriculum Correlation

Slade, Suzanne. *What if There Were No Bees? A Book About the Grassland Ecosystem.* Picture Window Books. Mankato, MN. 2011.

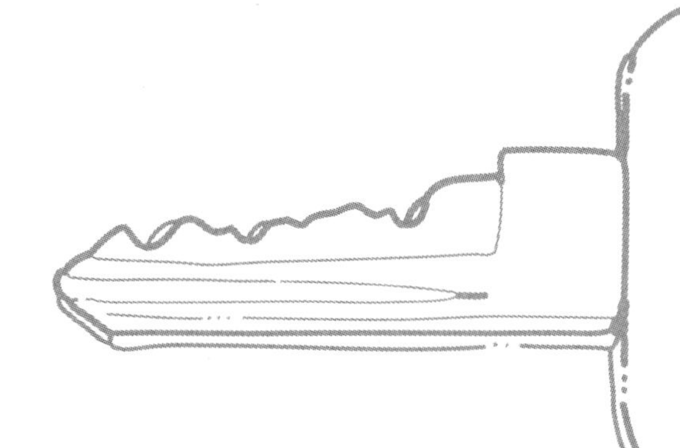

Key Question

How does a change in the food chain affect the ecosystem?

Learning Goal

Students will:

learn about how a change in the food chain affects an ecosystem by listening to and/or participating in a reader's theater.

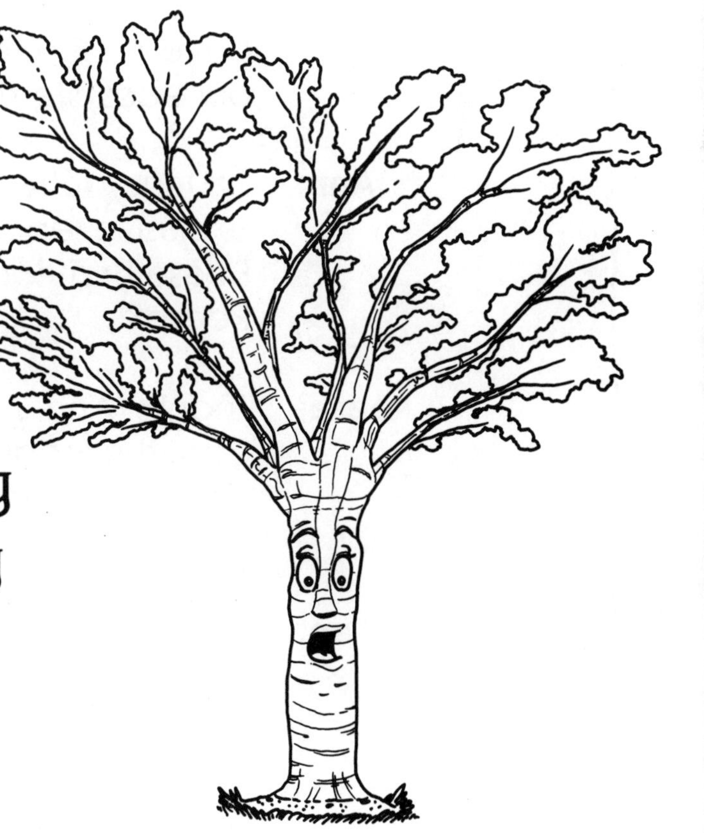

Crisis in the Greater Grasslands Ecosystem

Characters: Ann Chor, Rod Castor, Reese Urcher, Bizzy Bee, Apple Blossom, Cherry Blossom, Meeney Mouse, Sydney Snake, Harvey Hawk

Ann: Good evening, and thank you for joining us for the Greater Grasslands six o'clock newscast. I'm Ann Chor.

Rod: And I'm Rod Castor.

Ann: Well, as those of you who follow the news know, the situation has been going from bad to worse here in the Greater Grasslands ecosystem for the past six months.

Rod: Heck, Ann, it doesn't matter if you follow the news or not. There's no way anyone who lives here could have missed the changes that have happened in our ecosystem.

Ann: I guess you're right, Rod. Tonight, we have an exclusive report from our investigative reporter Reese Urcher on how this whole mess started. Reese?

Reese: Thanks, Ann. As you may recall, about six months ago, the local bee population in the Greater Grasslands ecosystem began to decline. At the time, I spoke to some of the few remaining bees.

Bizzy Bee: None of us is sure what's happening, but all I know is we're dropping like flies! Pesticides killed some of us, and others got sick and died. Those of us who are left are feeling lots of stress. I don't know how much longer we can hang on.

Reese: Well, as we now know, they didn't hang on much longer. Within a week of that interview, all of the bees in the Greater Grasslands ecosystem were gone. At first, we didn't notice any changes, but then, bad things started to happen. You may remember this interview from a few weeks later when I spoke to some of the local flora.

Apple Blossom: Normally at this time of year, our apple orchard is buzzing with bees. This year, there has been no buzz at all.

Cherry Blossom: It's been the same in our cherry orchards. When the bees come and drink our nectar, they also transfer our pollen from one flower to another. Without that pollination, the flowers won't turn into fruit. There will be no cherries on my tree.

Apple Blossom: There won't be any apples either. And it's not just us; many other local plants are also suffering. Most of our local fruits and vegetables are pollinated by bees.

Reese: Well, that shortage of food began to be felt in the food chain soon enough. We talked to Meeney Mouse shortly after our interview with the fruit trees.

Meeney Mouse: I eat lots of nuts, seeds, and berries at this time of year. Now I don't have enough to feed my family. Five of my cousins have already died of starvation, and my brother is talking about leaving Greater Grasslands to find food. I hate to leave my home, but if there's no food here, I may have no choice.

Reese: And it wasn't just the mice. Lots of other small animals and birds who rely on the plants pollinated by bees began to feel the pinch. Before long, larger animals were feeling the effects of the bees' deaths. About four months ago, I spoke to Sydney the snake about the changes.

Sydney Snake: Well, mice make up a large part of my diet, and about a week ago, I noticed that they were becoming harder to find. Now, I could stand to lose some weight, so at first, it was kind of nice. But now I'm getting worried. I haven't seen a mouse for three days, and my skin is starting to get baggy. I am having to spend lots more time looking for other things to eat, and that takes lots of energy! If the mice don't come back, I don't know what I'll do. There aren't enough toads around to feed my whole family and me.

Reese: Of course, the mice didn't come back, and it wasn't long before the hawks, owls, and foxes were starting to run out of food, too. This report was filed just last week when I interviewed some of our top predators.

Harvey Hawk: As a hawk, I'm at the top of the food chain. I have lots of choices of what to eat. Of course, I prefer rodents like mice and squirrels, but I have no objection to eating snakes, rabbits, or even bats. But let me tell you, food is getting hard to find! If things don't change soon, I'll be heading to greener pastures, if you know what I mean.

Reese: And we do know what he means. Since the bees stopped pollinating, many of our flowering plants have died. Because the flowers couldn't make seeds, there were no new plants to grow up and replace them. Instead of an ecosystem full of color, all we have left are a few grasses and shrubs. Our fruit trees are still growing, but without any fruit, they don't offer food to our animal population. Without food, animals died or moved away, and that has caused the chain reaction that has led to where we are today.

Rod: Wow, Reese. That was excellent reporting.

Ann: Absolutely! I had no idea all of these horrible changes to our ecosystem could be traced to the bees disappearing.

Reese: That's right, Ann. If I've learned anything from reporting this story it's that every part of the food chain is important. If something happens to one part, even a small one like the bees, the whole ecosystem suffers.

Rod: Well, that's it for tonight. We encourage you to tune in tomorrow when we discuss things we can do to get the bees back. Until then, I'm Rod Castor.

Ann: And I'm Ann Chor.

Ann & Rod: And this has been the Greater Grasslands news.

CHAIN REACTIONS
Ann
Chor

CHAIN REACTIONS
Rod
Castor

CHAIN REACTIONS
Reese
Urcher

CHAIN REACTIONS
Bizzy
Bee

CHAIN REACTIONS
Apple
Blossom

Cherry Blossom

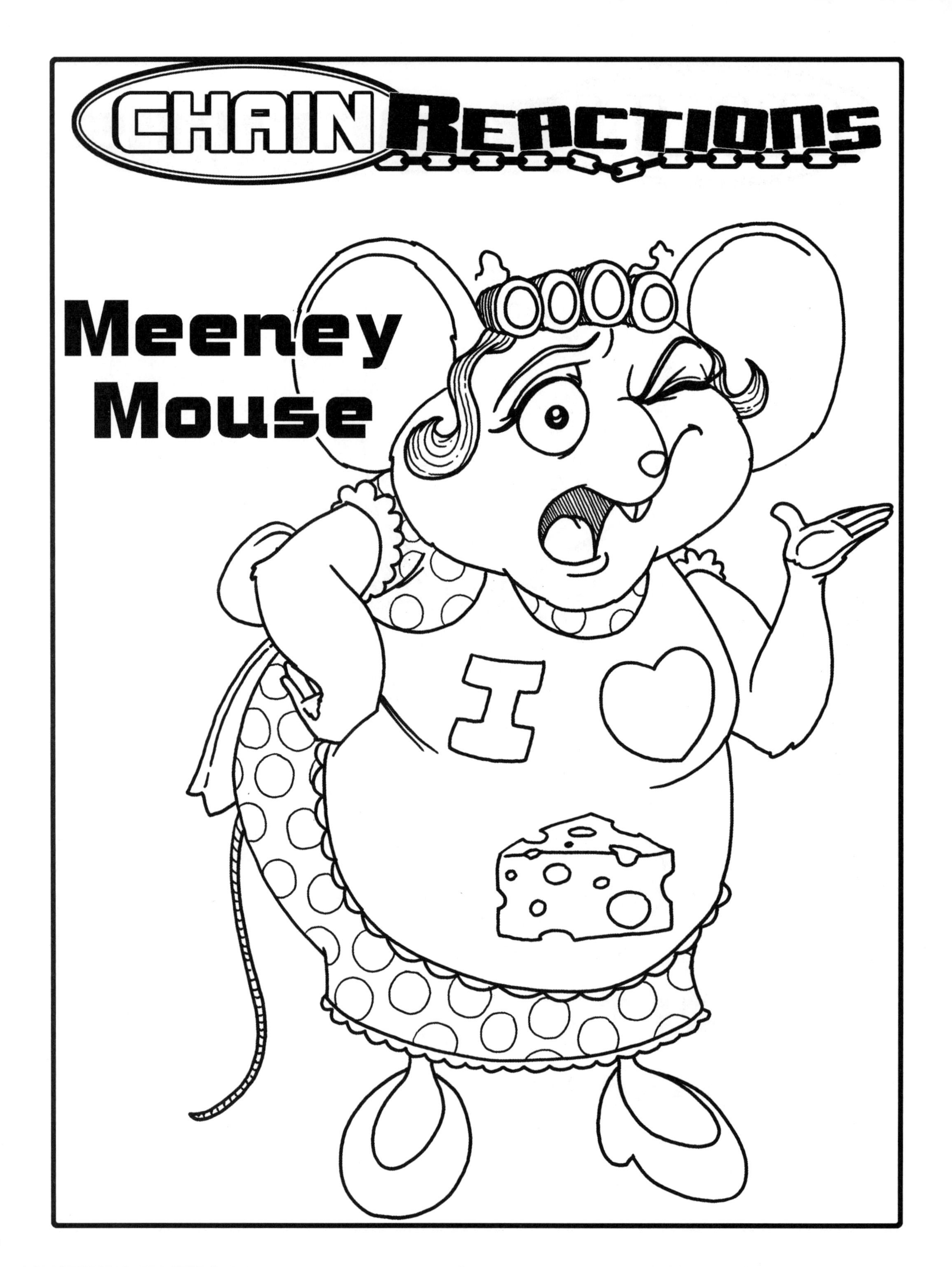
CHAIN REACTIONS
Meeney
Mouse
I

CHAIN REACTIONS
Sydney
Snake

CHAIN REACTIONS
Harvey
Hawk

CONNECTING LEARNING

Connecting Learning

1. What is one food chain in the grasslands ecosystem that you learned about in the reader's theater?
2. Where do the bees fit in this food chain?
3. What other food chains did you learn about from the book?
4. What would happen to the grassland ecosystem if there were no more bees?
5. What did you like most about this reader's theater?
6. What are you wondering now?

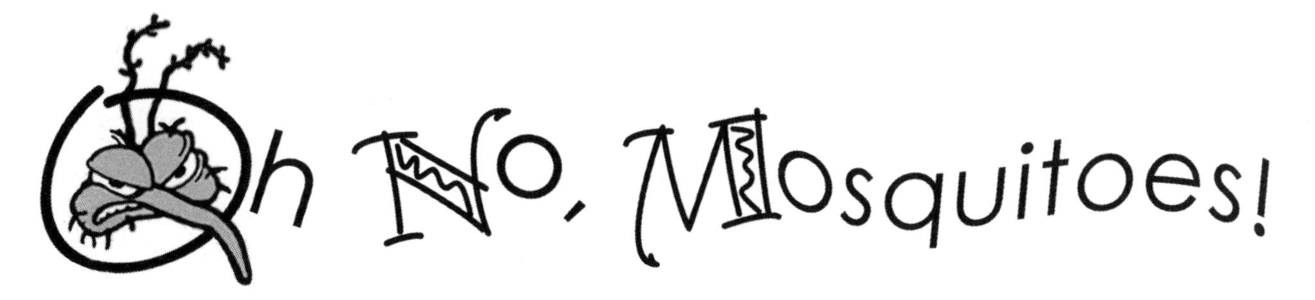

Topic
Food chains

Key Question
What happens when frogs are removed from the food chain?

Learning Goals
Students will:
- read about a pond ecosystem that turns from healthy to sick, and
- use frog and mosquito models to represent what happens when the number of frogs decreases in a pond ecosystem.

Guiding Documents
Project 2061 Benchmarks
- *Changes in an organism's habitat are sometimes beneficial to it and sometimes harmful.*
- *Some source of "energy" is needed for all organisms to stay alive and grow.*

NRC Standards
- *Organisms have basic needs. For example, animals need air, water, and food; plants require air, water, nutrients, and light. Organisms can survive only in environments in which their needs can be met. The world has many different environments, and distinct environments support the life of different types of organisms.*
- *All animals depend on plants. Some animals eat plants for food. Other animals eat animals that eat the plants.*
- *An organism's patterns of behavior are related to the nature of that organism's environment, including the kinds and numbers of other organisms present, the availability of food and resources, and the physical characteristics of the environment. When the environment changes, some plants and animals survive and reproduce, and others die or move to new locations.*

*Common Core State Standards for Mathematics**
- *Represent and solve problems involving multiplication and division. (3.OA)*
- *Use place value understanding and properties of operations to perform multi-digit arithmetic. (3.NBT, 4.NBT)*
- *Use the four operations with whole numbers to solve problems. (4.OA)*

Math
Whole number operations
  multiplication

Science
Life science
  food chains
  ecosystems
    ponds

Integrated Processes
Observing
Comparing and contrasting
Recording data
Interpreting data
Relating

Materials
For each student:
paper plate, 6-inch (see *Management 3)*
glue stick
red yarn, 30 cm
tape
swarm of mosquitoes picture
scissors
student pages

For the class:
green copy paper
The Pond in the Park story, included
sticky notes, 3-inch (see *Management 7)*

Background Information

A healthy pond ecosystem must have a delicate balance of non-polluted oxygenated water, plants, and animals. All levels of the pond—above the surface, on the surface, and below the surface—must be healthy and balanced. The pond ecosystem has its own food web with some of the players as permanent residents, while others may be temporary guests (e.g., some may move in to feed and leave to find shelter).

Producers found in a pond ecosystem include the plants at the edge of the pond, floating plants, submerged plants, and plants that are rooted in shallow areas but have stems and leaves above the surface. Microscopic algae are also producers in the pond ecosystem. Like other producers, they get their energy from the sun.

Consumers range in size from microscopic herbivores and carnivores to larger organisms such as turtles, frogs, and herons.

Dead plants and animals, along with animal waste, form detritus on the bottom of the pond. Decomposers break down the detritus into the nutrient-rich materials that the producers use.

This simplified activity explores what happens to a portion of a pond's food chain when the population of frogs is reduced. The food chain looks at mosquitoes as the source of food (prey) for the frogs and the heron as the predator of the frog. Students will find that when one link of a food chain is broken, the entire chain is weakened.

Two key elements are embedded in this lesson:

1) Some frogs (e.g., the dwarf puddle frog) can eat 100 mosquitoes in one night.
2) Frogs absorb moisture through their skin. If the water has been polluted, that pollution is absorbed into the frogs' bodies. This is a current environmental concern. Many frogs have been killed or deformed because of the runoff of chemical herbicides used on farms and yards into bodies of water.

Management

1. Each student will need to make a frog and mosquito model.
2. Cut the yarn into 30-cm lengths. Each student will need a length.
3. Locate an area in which all the students can form a circle and sit down on the floor. This will form the outline of the pond.
4. Flimsy 6-inch paper plates work well for this activity.
5. Copy the eyes and the legs on green copy paper. Each student needs one set.
6. Beginning with one, write a number on each sticky note. Make enough to represent the number of students in the class.
7. Prepare a frog and mosquito model beforehand so that students can see a finished product.

Procedure

1. Ask the students to describe a pond ecosystem, being sure they include both living and nonliving things. [plants, fish, frogs, mud, water, trees, bugs, etc.]
2. Tell them that they are going to learn what happens when the pond ecosystem becomes unhealthy.
3. Distribute materials for making the frog and mosquito models. Allow time for students to follow the directions to cut out and assemble the models. Give each student a sticky note with a number on it. Direct them to place the sticky notes on the backs of their frogs.
4. Invite the students to bring their frogs to the area you have selected. Tell them to line up numerically by the numbers on their frogs. Have them hold hands to form a circle. Once the circle is established, have them release hands and sit down, placing their frogs and mosquitoes in front of them. Tell the students that they are sitting around the edge of the pond.
5. Read the story of *The Pond in the Park* to them.
6. Ask the students to retell the story of the pond. Tell them that they are going to see what happened when the frogs started dying out.
7. Ask students how many frogs are around their pond. [the same as the number of students in the class] Tell the students that these frogs eat 100 mosquitoes each day. Let students skip count by 100s to see how many mosquitoes the frogs at the pond ate (e.g., 24 frogs x 100 = 2400 mosquitoes). Show students how to pull on the yarn so that the mosquitoes go to the frogs' mouths.
8. Have students extend the yarn so the mosquitoes are ready for the next round. Question students as to why the frogs were dying. [The polluted water killed them.] Tell them that at first not many frogs died. To show that just a few died in the pond, direct everyone who has a frog with a number that is a multiple of five to take their frog out of the pond. They must leave the mosquitoes behind. If necessary, count by fives so that students understand which frogs are taken out of the pond.
9. As a class, count the number of frogs that were removed. Remind the students that each of these frogs would have eaten 100 insects. Have all the remaining frogs eat their mosquitoes. Count the mosquitoes that are left in the pond (e.g., if 5 frogs have been removed, 5 x 100 = 500 mosquitoes are in the pond).
10. Direct the students to put all the frogs back and to extend the yarn so the mosquitoes are ready for the next round.
11. Continue the scenario by saying that more frogs died. This time have all the even numbered frogs removed, leaving their mosquitoes behind. Talk about how the population of mosquitoes has increased and could cause problems. Ask the students what problems mosquitoes can cause. [bites and diseases]
12. Have students pick up their frogs and mosquitoes and return to their desks. Distribute the student page and direct students to complete it.
13. End with a discussion about the impact of losing links in the food chain.

Connecting Learning

1. Name some living things in a pond ecosystem. [frogs, fish, insects, ducks, water plants, grasses, etc.]
2. Name some nonliving things in the pond ecosystem. [rocks, mud, water]
3. What animals found around ponds do frogs eat? [mosquitoes and other insects, spiders, worms, small fish]
4. What animals found around ponds eat frogs? [herons, hawks, snakes, lizards, birds, other small animals]
5. What happens to the ecosystem if the number of frogs goes down a lot? [the ecosystem becomes imbalanced, insect populations increase, etc.]
6. How did the herons respond to the change in the pond environment? [They moved to other locations.]
7. What are you wondering now?

Internet Connections

Ecokids

http://www.ecokids.ca/pub/eco_info/topics/frogs/chain_reaction/play_chainreaction.cfm

This is an interactive game in which students place the elements of a food chain in order. There are two food chain options, one for a forest food chain (more appropriate for this activity) and one of a northern food chain.

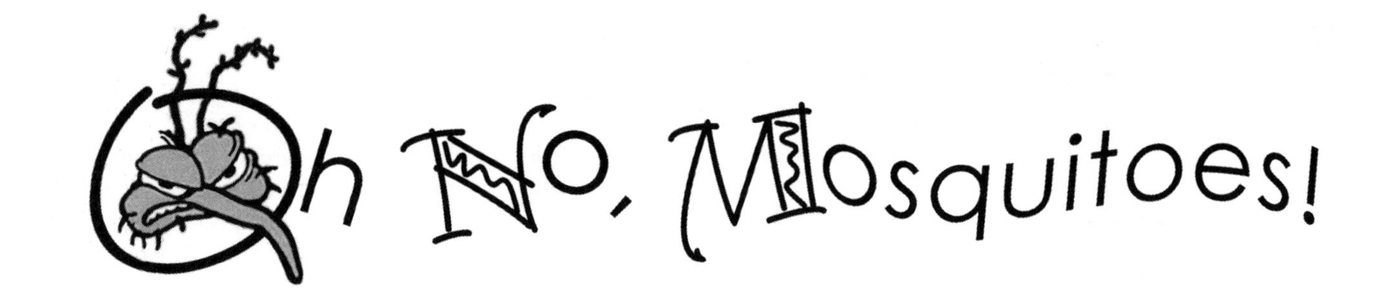

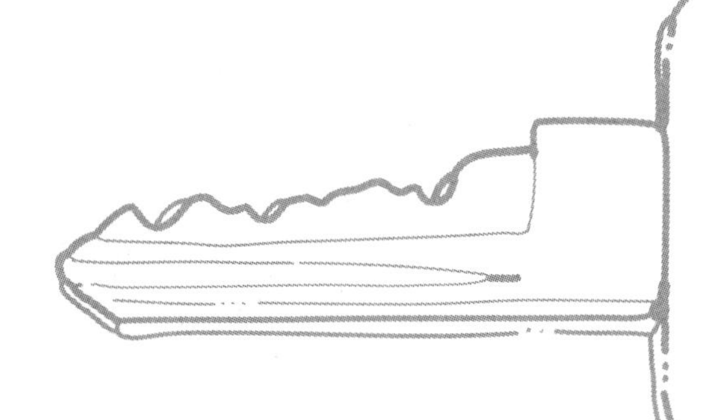

Key Question

What happens when frogs are removed from the food chain?

Learning Goals

- read about a pond ecosystem that turns from healthy to sick, and
- use frog and mosquito models to represent what happens when the number of frogs decreases in a pond ecosystem.

Oh No, Mosquitoes!

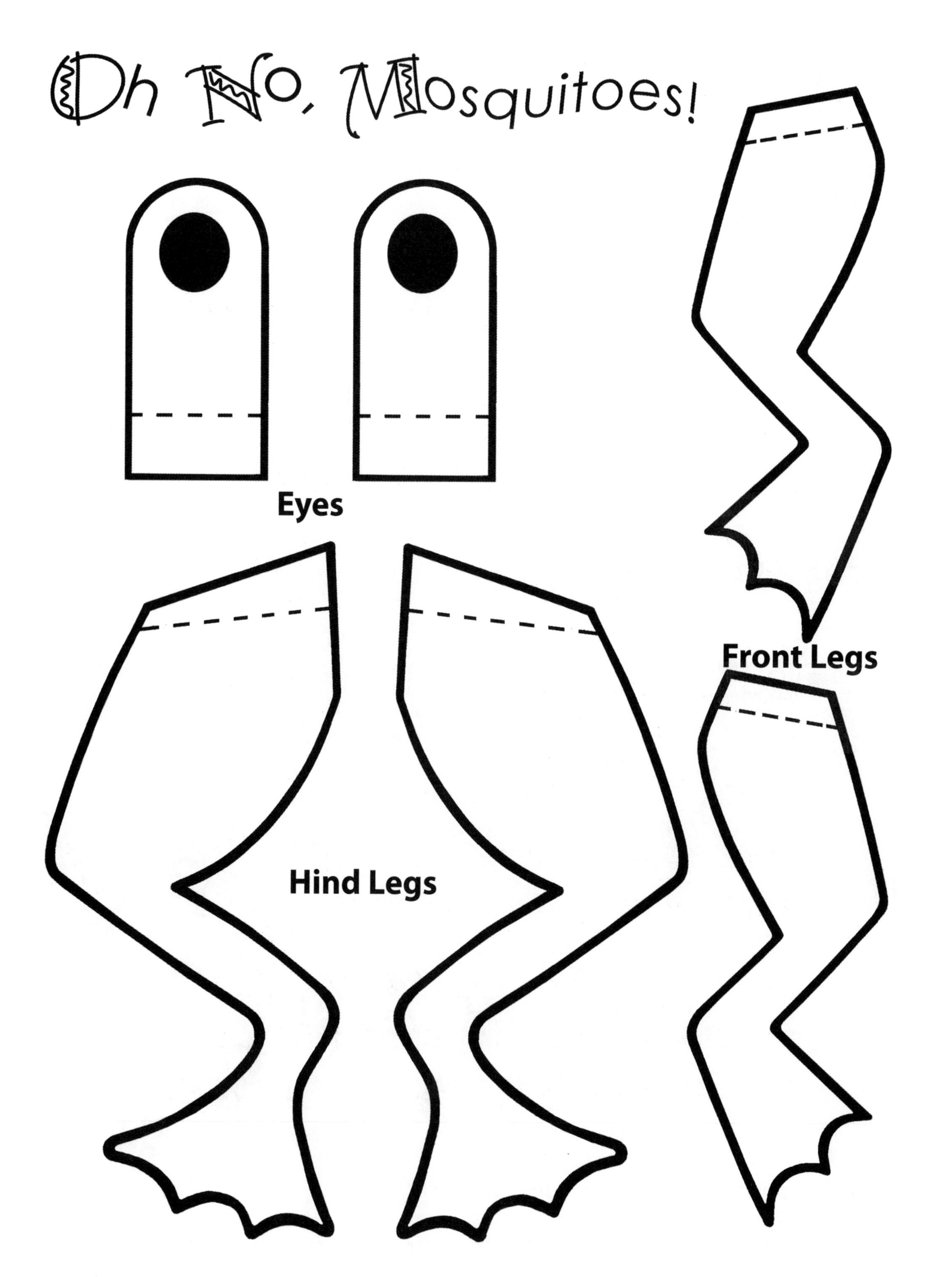

Oh No, Mosquitoes!

1. Fold the paper plate in half. Cut a small triangle out of the center of the fold.

2. Cut out the eyes, fold, and glue them onto the top of your folded paper plate.

3. Cut out the legs, fold, and glue them to the bottom of the folded paper plate.

4. Tape the swarm of mosquitoes to one end of the yarn. Thread the yarn through the mouth of the frog and through the hole you cut in the fold of the paper plate.

5. You can pull on the yarn to make the frog eat the mosquitoes.

The Pond in the Park

There was once a beautiful pond in the park. Tall, shady trees surrounded the pond. Lots of green grasses and flowering plants flourished. Small animals and birds used the grasses for shelter.

Ducks were seen swimming in the clean water. They were eating pond plants. Herons were eating minnows, frogs, and fish. Frogs were croaking and catching mosquitoes. Life was good at the pond. Food was plentiful.

People wanted to live near the pond. They built houses nearby. They put in beautiful lawns and gardens.

It was pretty, but not everything was perfect. The frogs started dying.

Because there were not as many frogs, the herons did not have enough food. The herons flew off to other ponds. There were more and more mosquitoes because there were not enough frogs to eat them.

The mosquitoes bit the children. The children no longer wanted to play in the park. It was no longer the wonderful place it once had been.

Why had the frogs died? Scientists had to help find the problem. They studied both the frogs and water. They found out that the water was polluted. People had used fertilizers on the park grass and on their lawns. They wanted to have beautiful green lawns. Fertilizers help make lawns nice and green.

People had also used herbicides. Herbicides are chemicals that kill weeds. No one wanted weeds in the park or in their lawns.

When it rained or when the people watered their grass, some of the chemicals were washed into the pond. The water became polluted from the chemicals.

How did the polluted water harm the frogs? Frogs don't drink water like we do. They absorb moisture from the water. The frogs had absorbed the polluted water. The polluted water harmed them and their babies.

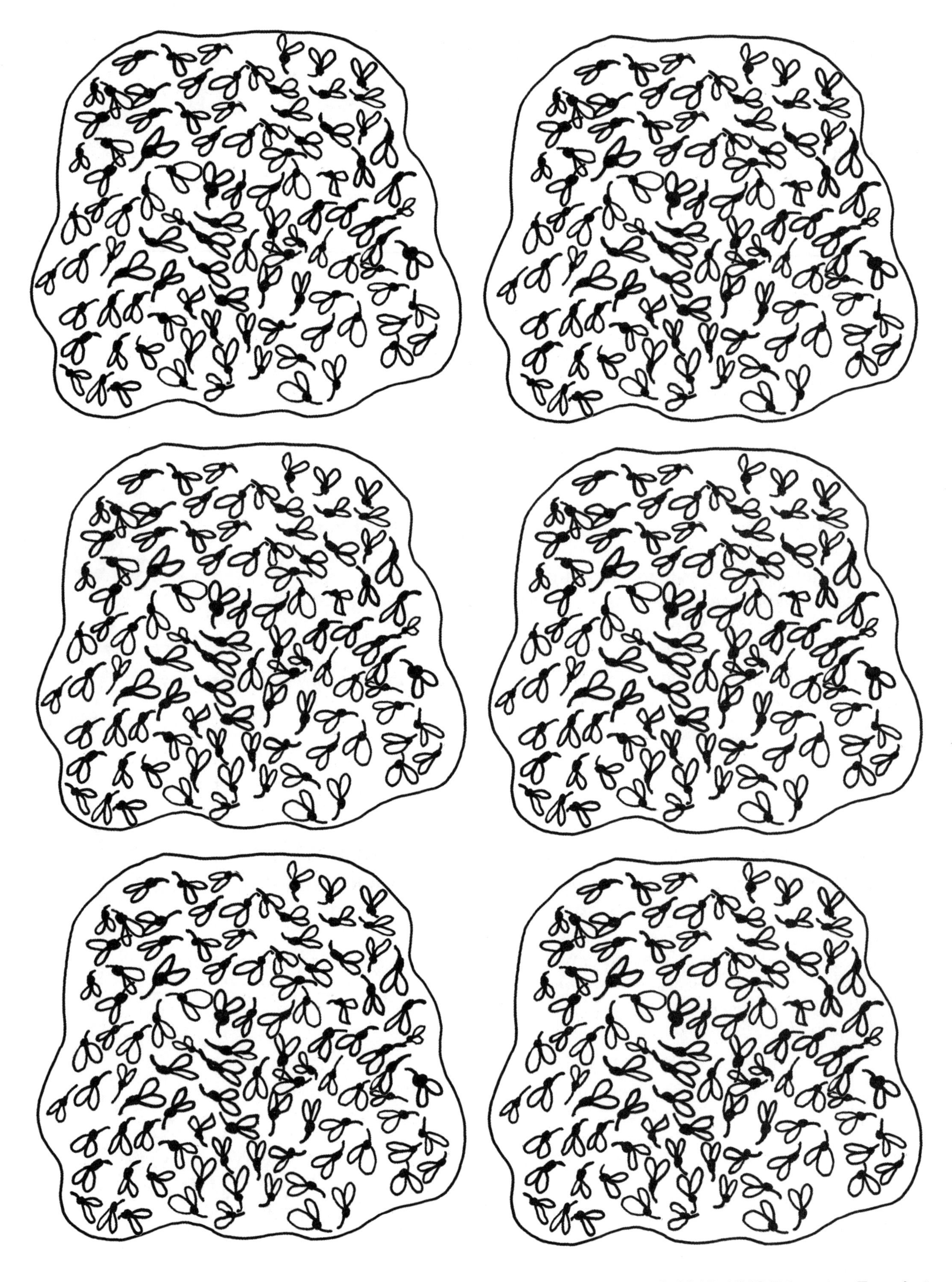

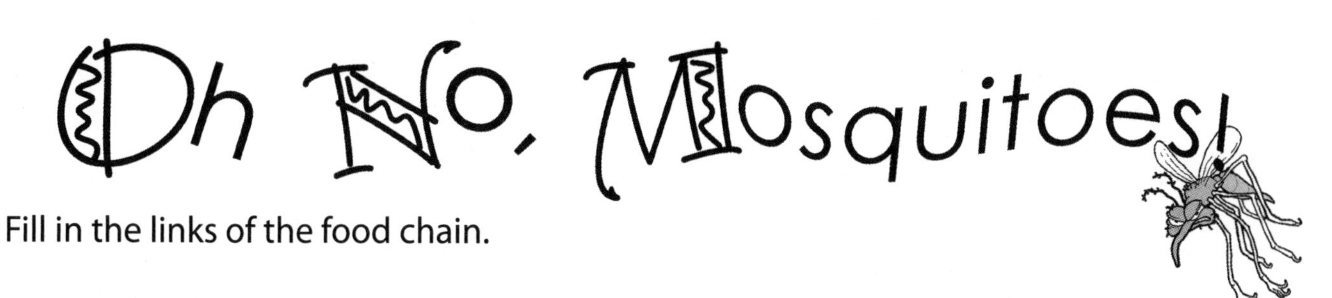

Fill in the links of the food chain.

____________________ → Frogs → ____________________

Suppose the large frogs in this pond each eat 100 mosquitoes. The smaller frogs each eat 50 mosquitoes each night. How many mosquitoes do both groups eat each night? Show your work.

Four large frogs and three smaller ones die because of polluted water. How many mosquitoes do not get eaten that night? Show your work.

If all the large frogs die, how many mosquitoes will not be eaten each night?

How many mosquitoes will be eaten by the smaller frogs?

Write two ways in which frogs are important to the pond ecosystem.

CONNECTING LEARNING

Connecting Learning

1. Name some living things in a pond ecosystem.
2. Name some nonliving things in the pond ecosystem.
3. What animals found around ponds do frogs eat?
4. What animals found around ponds eat frogs?
5. What happens to the ecosystem if the number of frogs goes down a lot?
6. How did the herons respond to the change in the pond environment?
7. What are you wondering now?

Topic
Food webs

Key Questions
1. How does energy flow through a food web?
2. How will a change to the ecosystem affect the food web?

Learning Goals
Students will:
- use clue cards to identify how plants and animals are connected in a food web,
- trace the flow of energy through the food web, and
- predict what would happen to the food web if certain changes in the ecosystem took place.

Guiding Documents
Project 2061 Benchmarks
- *Some source of "energy" is needed for all organisms to stay alive and grow.*
- *In something that consists of many parts, the parts usually influence one another.*
- *Seeing how a model works after changes are made to it may suggest how the real thing would work if the same were done to it.*

NRC Standards
- *All animals depend on plants. Some animals eat plants for food. Other animals eat animals that eat the plants.*
- *Organisms have basic needs. For example, animals need air, water, and food; plants require air, water, nutrients, and light. Organisms can survive only in environments, and distinct environments support the life of different types of organisms.*

Science
Life science
  food webs
  ecosystems

Integrated Processes
Observing
Relating
Predicting
Identifying

Materials
12" x 18" construction paper
Scissors
Glue sticks
Colored pencils
Clue cards (see *Management 2)*
Student pages

Background Information
Plants and animals need energy to live and grow. The source of energy for life on Earth is the sun. The sun provides the energy necessary for green plants to make their own food. The plants, in turn, provide energy to the animals that eat them. These animals provide energy to other animals that eat them, and the transfer of energy continues.

These transfers of energy from plant to animal to animal are known as food chains—the mouse eats the grain; the snake eats the mouse; the hawk eats the snake. Since most animals eat more than one thing, food chains intersect forming food webs. These food webs are more accurate representations of how animals within an ecosystem depend upon each other. A change at any point within a food web will have implications throughout. If a fire wipes out the vegetation, it is not only the plant-eaters who suffer but also the animals that eat the plant eaters.

Management
1. Students should work together in groups of four so that each member of the group is responsible for one clue card. Where groups of four are not possible, students can pair up and share clue cards, or one student can be responsible for two clue cards.
2. Copy the clue cards onto card stock and cut them apart ahead of time. Make one set for each group of four students.
3. Plan to display the sample food web page using a projection device. You will need to be able to draw in arrows indicating the flow of energy in the food web.
4. This activity is divided into two parts. In *Part One,* students construct a model food web for a forest ecosystem and trace the flow of energy through the food web. In *Part Two,* they predict the effects of changes to the ecosystem on the food web. These parts can be done on different days, if desired.

Procedure

Part One

1. Ask students to describe what they have learned about food chains. Challenge them to identify the difference between a food chain and a food web. [A food web is many food chains that intersect.]
2. Display the sample food web page and ask the first *Key Question:* "How does energy flow through a food web?" Solicit student responses, and challenge them to identify the connections in the displayed food web based on what they know about the animals listed.
3. Draw in the arrows that illustrate the flow of energy in the food web. The plants give energy to the grasshopper and the vole. The vole gives energy to the snake and the hawk. The grasshopper gives energy to the toad. The toad gives energy to the snake, and the snake gives energy to the hawk.

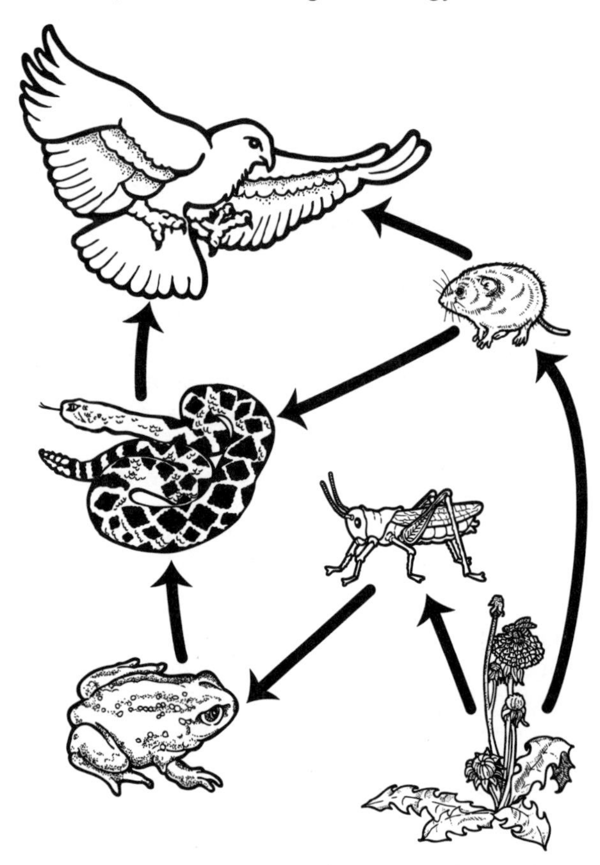

4. Explain to students that they will be working in groups to create their own model food webs like the one you just did using animals in a forest ecosystem. They will use logic clues to determine how the animals and plants in the food web are connected.
5. Divide the students into groups of four and give each group a piece of 12" x 18" construction paper, the page of plant and animal cutouts, and scissors.
6. Instruct groups to cut out the plants and animals by rough cutting around their pictures and to arrange them on the construction paper so that they are spread out. Tell students that they will not be allowed to glue the plants and animals down until they have determined all of the connections, because they may want to put them in different places to make the connections easier to see.
7. Distribute one set of clue cards to each group. Explain that each student in the group is responsible for one clue card. They will take turns reading the clues on their cards aloud to the group and will work together to use the information to determine the connections in the food web.
8. Allow time for groups to determine the connections between the plants and animals. Students may want to use pencil to lightly indicate connections on the construction paper.
9. When groups are ready, distribute the glue sticks and colored pencils. Have them glue the pictures in place and draw the arrows between the plants and animals. Be sure they draw the arrows showing the direction that the energy flows in the food chain (from the plant/animal being eaten to the animal doing the eating).
10. Tell them that they must go through their clue cards one more time and double check that they have addressed each clue to be sure that all of the connections are shown in their food webs.
11. When all groups have completed their food webs, ask the students if they think anything is missing from their diagrams. If no one mentions it, point out that they do not show where the blueberry bushes get their food.
12. Invite suggestions for what should be done to show the complete flow of energy through the food chain. [Add the sun and connect it to the berry bush.]

Part Two

1. Have students get out their food web diagrams. Discuss the ecosystem in which this food web appears—the forest.
2. Ask the second *Key Question:* "How will a change to the ecosystem affect the food web?" Invite students to brainstorm some changes that could take place in a forest ecosystem. [Forest fire, logging (removal of trees), disease that kills certain plants and/or animals, pollution in the water sources, drought, etc.]
3. Tell students that a forest fire has come through and burned three-quarters of the plants in the forest. Ask them what would happen to the forest food web as a result. [Plants would be harder to find for the mice, deer, and birds. They might die or move to where they could get more food. With fewer prey, the fox and the wolf populations would decline.]

4. Continue to give different scenarios, including those that may have a positive effect, and have students predict the changes in the food web that would result.

Connecting Learning

1. What is the difference between a food chain and a food web? [A food chain is a series of plants and animals that each depends on the next as a source of food. A food web is many food chains that intersect.]
2. What were the connections in your forest ecosystem food web? (see *Solutions.)*
3. Did any of the connections surprise you? Explain.
4. How do changes to the forest ecosystem affect the food web?
5. What are you wondering now?

Curriculum Correlation

Slade, Suzanne. *What if There Were No Gray Wolves? A Book About the Temperate Forest Ecosystem.* Picture Window Books. Mankato, MN. 2011.

Solutions

The connections in the food web are shown here. Students may have arranged the pictures differently, but the connections should be the same.

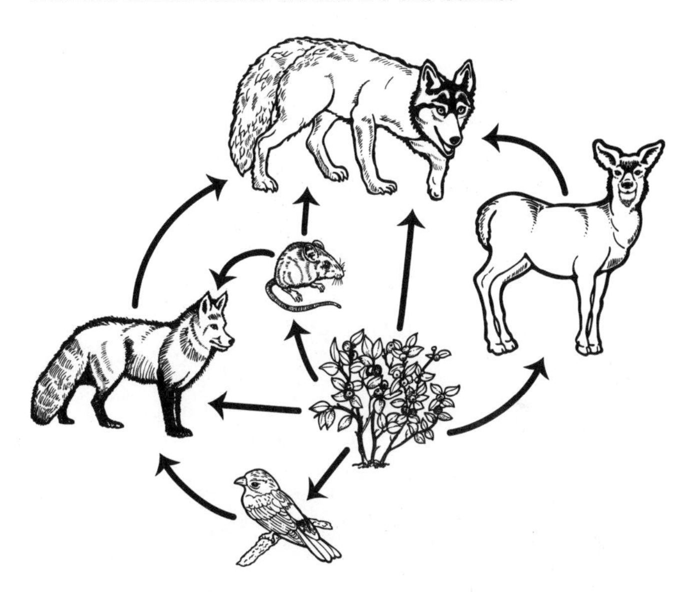

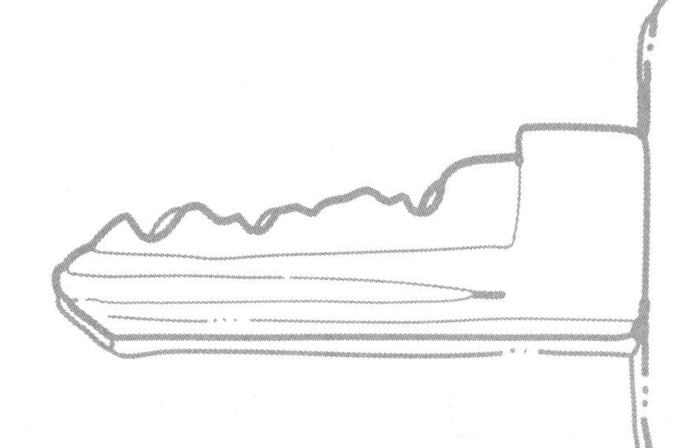

Key Questions

1. How does energy flow through a food web?
2. How will a change to the ecosystem affect the food web?

Learning Goals

Students will:

- use clue cards to identify how plants and animals are connected in a food web,
- trace the flow of energy through the food web, and
- predict what would happen to the food web if certain changes in the ecosystem took place.

Web Work

Sample Food Web

Vole

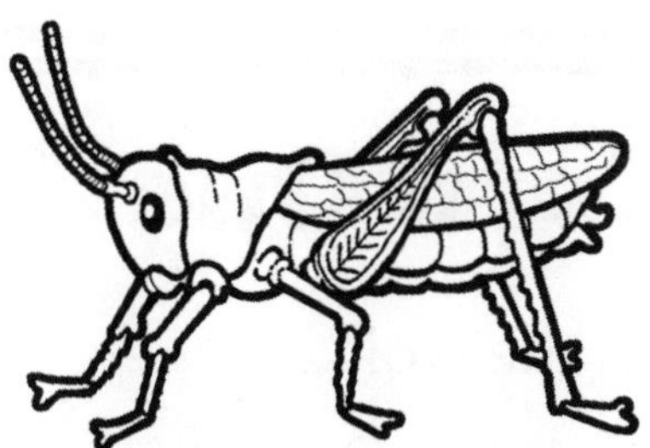

Web Work Clue Cards

Clue Card One

- Nothing eats the wolf.
- Five animals rely on the berry bush for food.

Clue Card Two

- Wolves do not eat birds.
- Only one animal eats deer.
- The mouse is food for two animals.

Clue Card Three

- The bird and the mouse are both eaten by the same animal.
- The fox only eats smaller animals.

Clue Card Four

- The deer, the mouse, and the bird eat only one thing.
- The wolf eats three animals.

Clue Card One

- Nothing eats the wolf.
- Five animals rely on the berry bush for food.

Clue Card Two

- Wolves do not eat birds.
- Only one animal eats deer.
- The mouse is food for two animals.

Clue Card Three

- The bird and the mouse are both eaten by the same animal.
- The fox only eats smaller animals.

Clue Card Four

- The deer, the mouse, and the bird eat only one thing.
- The wolf eats three animals.

Web Work

Cut apart the plants and animals. Arrange them on your paper to show the connections.

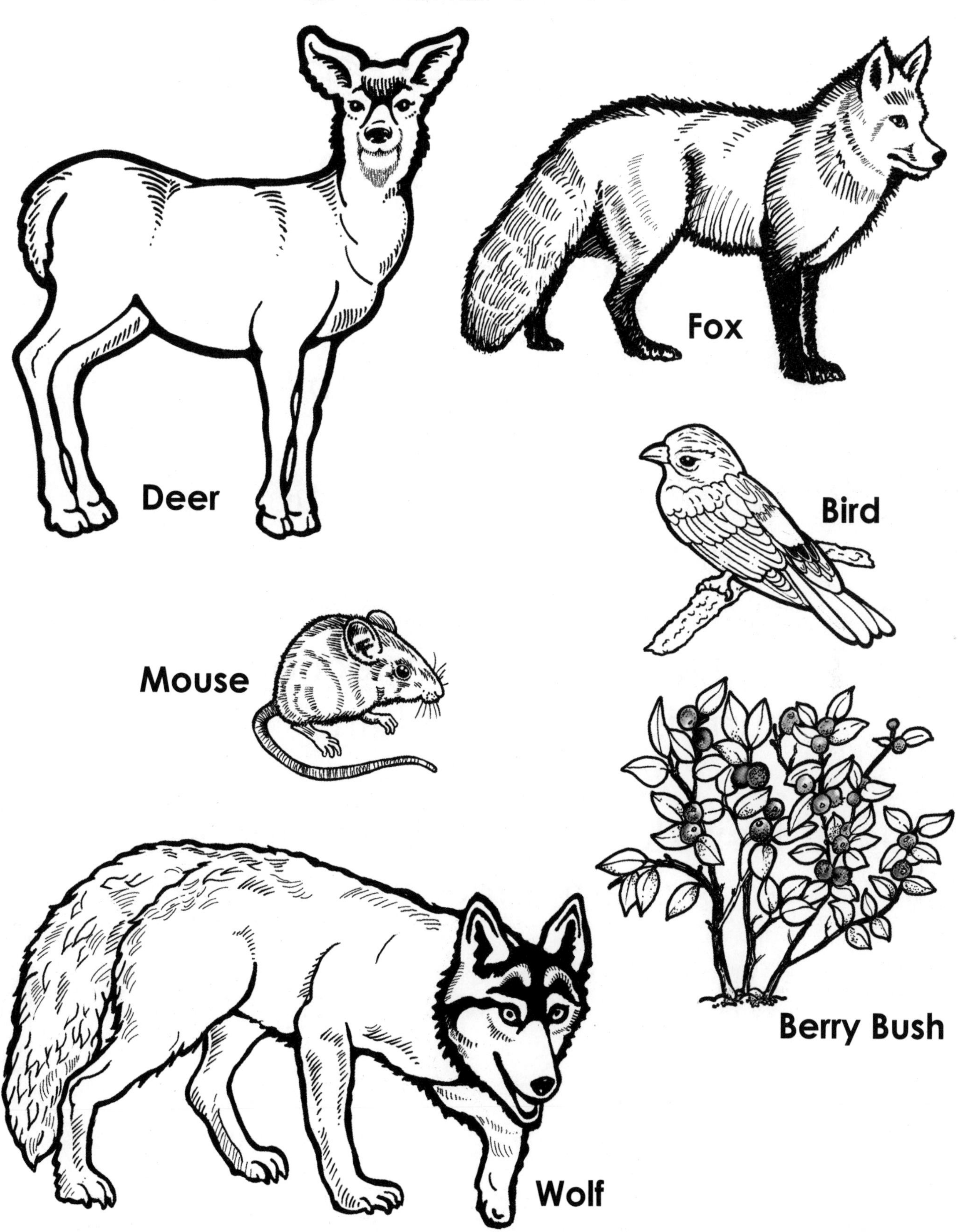

Web Work

CONNECTING LEARNING

Connecting Learning

1. What is the difference between a food chain and a food web?
2. What were the connections in your forest ecosystem food web?
3. Did any of the connections surprise you? Explain.
4. How do changes to the forest ecosystem affect the food web?
5. What are you wondering now?

Topic
Food webs

Key Questions
1. What is an ocean habitat like?
2. How do plants and animals in the ocean depend on each other?

Learning Goals
Students will:
- identify the ocean as a habitat,
- learn some of the typical plants and animals in an ocean habitat, and
- discover how these plants and animals depend on each other for survival.

Guiding Documents
Project 2061 Benchmarks
- *Some source of "energy" is needed for all organisms to stay alive and grow.*
- *In something that consists of many parts, the parts usually influence one another.*
- *Seeing how a model works after changes are made to it may suggest how the real thing would work if the same were done to it.*

NRC Standards
- *All animals depend on plants. Some animals eat plants for food. Other animals eat animals that eat the plants.*
- *Organisms have basic needs. For example, animals need air, water, and food; plants require air, water, nutrients, and light. Organisms can survive only in environments, and distinct environments support the life of different types of organisms.*

Science
Life science
food webs
habitats
interdependence

Integrated Processes
Observing
Communicating
Collecting and recording data
Applying

Materials
Plant and animal fact cards
String cut in 1-2 meter lengths
Tape
Scissors

Background Information
In every environment, the organisms present interact with each other and depend on each other in a variety of ways. One of the ways to examine these interrelationships is through the use of food webs. Food webs show the connections between animals and what they eat (or what eats them). Food chains and webs contain plants *(producers)*, which use sunlight and inorganic materials to produce the organic compounds that become food and nutrients for other organisms—the *consumers*. The animals that feed on plants are called *primary consumers*; animals that eat other animals are *secondary consumers*. *Scavengers* then feed on dead organisms, while *decomposers* break down nonliving organic matter into materials that again are available to enter the food chain as *nutrients*. Nutrients in an ocean habitat tend to settle at the bottom of the ocean.

This activity includes plants in the food web and looks specifically at an ocean habitat.

Management
1. This activity is divided into two parts. In the first part, students will complete their plant and animal fact cards by filling in the "things that eat me" section. In the second part, they will create a simulated food web and see what happens when one or more parts of the web is removed.
2. It is best to copy the animal fact cards onto card stock for extended use.
3. Before doing *Part Two*, cut string into one- to two-meter lengths. Depending on your class size, you will need between 45 and 55 separate pieces.
4. There are 21 plant and animal cards provided. If you need more than that for your class, you can either add your own animals or make multiple copies of some of the cards. If you have fewer than 21 students, you will need to remove some of the cards. It is recommended that you remove only animals with minimal connections, such as jellyfish.

5. The information provided on the plant and animal fact cards is intentionally generalized for the sake of simplicity. There are many things that each animal may eat that are not listed. There are also things listed that may be eaten only by certain species of the animals.

Procedure

Part One

1. Ask students if they can think of some of the different kinds of plants and animals that would live in an ocean habitat. Write their suggestions on the board.
2. Discuss the location of any oceans near you, and have students who may have been there share their memories of the plant and animal life.
3. Tell students that they are going to do an activity where they take on the role of a plant or animal that lives in an ocean habitat. Give each student one plant or animal fact card.
4. Explain that this card tells them who they are and what they eat. Their job is to find the people who have the fact cards for every plant and/or animal that they eat. When they find these fact cards, they are to write the name of their animal on that person's card under the heading: *Here are some things that eat me.*
5. Tell students that the animal names on their cards might not be specific. They may have to use their problem-solving skills to find some of the animals they eat. (For example, the killer whale eats "fish," but there are no "fish" animal cards. Students must figure out that herring and salmon are two kinds of fish.)
6. Allow the class time to circulate and find every plant and/or animal on their lists. Monitor their progress and check for errors and/or omissions.
7. When they are done, discuss what they learned and discovered by doing this activity.

Part Two

1. Have all the students stand in an open area, holding their fact cards.
2. Give every few students a bundle of pre-cut string and a roll of tape. Instruct them to tape one piece of string next to the name of every animal under their "things that eat me" list.
3. Once all students are done taping, have them hold their fact cards in one hand and use the other to hold all of the strings that go to the name of their plant or animal on someone else's card.
4. Discuss how the different plants and animals are connected. Explain that this is a model of a food web. They are showing some of the ways that these plants and animals depend on each other for food.
5. Ask students what they think would happen if one of the plants or animals were taken out of the food web.
6. Tell a story that removes students from the food web, one at a time. (e.g., *The ocean was full of fishermen and they have caught all of the salmon.)* Start with plants and low-level consumers so that each removal has the maximum effect on the food web.
7. Explain to students that each time a plant or animal is removed, anyone holding strings connected to them must let go. Tell them that when any animal has no more strings to hold on to (food sources), it is dead and must also leave the food web.
8. See how many things have to be removed before all of the animals are dead. Repeat the process as desired, removing different animals in a different order, and discuss what was learned.

Connecting Learning

Part One

1. Describe an ocean habitat.
2. What kinds of plants and animals live in that habitat?
3. Could any of the animals that we studied live in any other habitat? Explain your thinking.
4. What are some of the differences between the animals and what they eat? [some eat plants, some eat animals, some eat lots of things, some eat only a few things]
5. What do the plants eat? [Nothing. They make their own food using energy from the sun.]

Part Two

1. How was our model like a real food web? [It showed how plants and animals depend on each other for food.] How was it different? [A real food web starts with the sun. There would be many more animals and connections in a real food web.]
2. What happened when plants and animals were taken out of the food web? [Other animals were affected.]
3. Which animals died first (without being removed by the teacher)? [those that had only a few food sources]
4. Which animals survived the longest? [those that had the most sources of food]
5. Which plants or animals had the biggest impact when they were removed? [producers, low-level consumers] Why? [lots of other animals eat them]
6. Which plants or animals had the least impact when they were removed? [top-level consumers] Why? [nothing eats them]
7. How do you think people fit in to our food web?
8. What are you wondering now?

Extensions

1. Take a field trip to a nearby ocean or aquarium. Look for some of the animals that you studied in this activity, and make a list of others that you observe.
2. Do research and add additional animals to the food web.
3. Make food webs for other habitats.

Internet Connection

Monterey Bay Aquarium
http://www.montereybayaquarium.org/
Click on the "Animal Guide" link under the "Animals & Activities" tab to access an extensive database of sea animal pictures and facts.

Enature
http://www.enature.com
Field guide information on North American plants and animals can be found on this website. You can also search by zip code for plants and animals in your area.

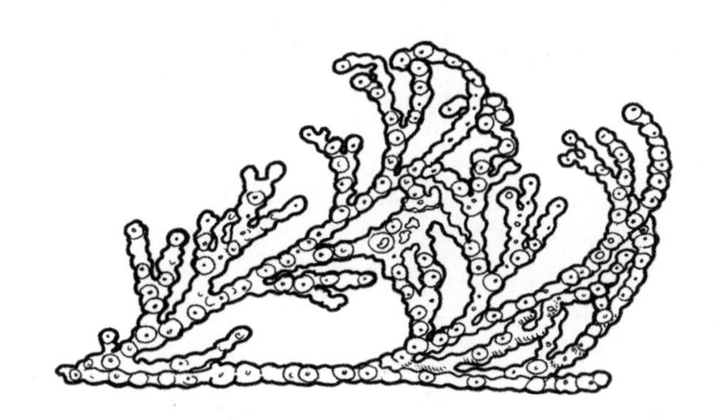

Key Questions

1. What is an ocean habitat like?
2. How do plants and animals in the ocean depend on each other?

Learning Goals

Students will:

- identify the ocean as a habitat,
- learn some of the typical plants and animals in an ocean habitat, and
- discover how these plants and animals depend on each other for survival.

I am a killer whale.
I eat fish, seals, squid, sea birds, dolphins, and sea turtles.
No animals eat me.

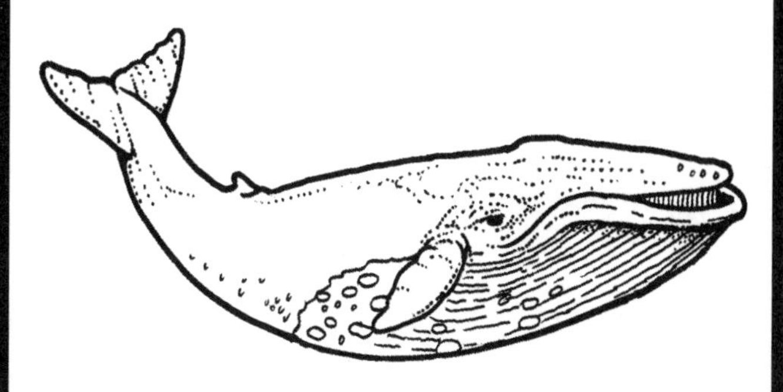

I am a blue whale.
I eat krill.
No animals eat me.

I am a seagull.
I eat herring, clams, sea stars, crabs, baby turtles, and the eggs of other birds.

Here are some things that eat me:

____________________ ____________________

____________________ ____________________

____________________ ____________________

____________________ ____________________

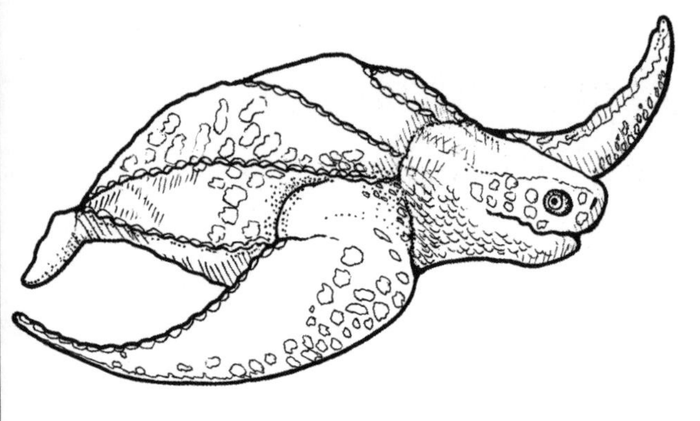

I am a sea turtle.
I eat jellyfish, crabs, shrimp, small fish, sea grass, and algae.

Here are some things that eat me:

I am a bottlenose dolphin.
I eat fish and squid.

Here are some things that eat me:

I am a harbor seal.
I eat fish, squid, clams, and krill.

Here are some things that eat me:

I am a salmon.
I eat herring, squid, and shrimp.

Here are some things that eat me:

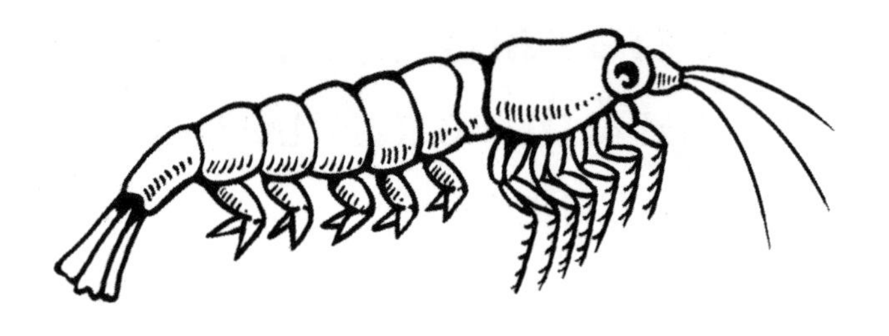

I am a krill.
I eat phytoplankton
and zooplankton.

Here are some things that eat me:

________________ ________________

________________ ________________

________________ ________________

________________ ________________

I am a squid.
I eat fish, shrimp, and
other squid.

Here are some things that eat me:

________________ ________________

________________ ________________

________________ ________________

________________ ________________

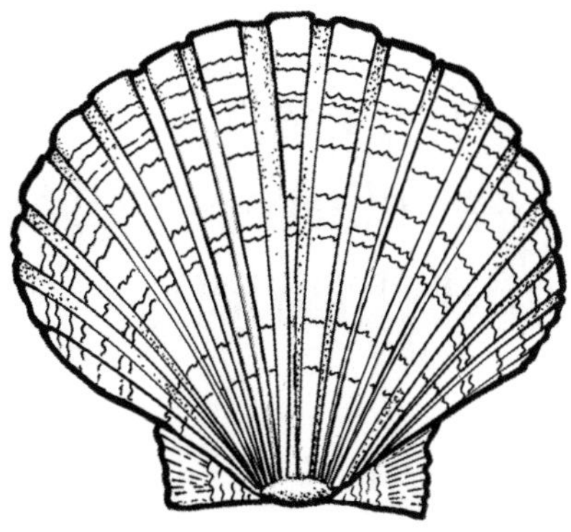

I am a scallop.
I eat phytoplankton
and algae.

Here are some things that eat me:

______________________ ______________________

______________________ ______________________

______________________ ______________________

______________________ ______________________

I am a brown pelican.
I eat herring.

Here are some things that eat me:

______________________ ______________________

______________________ ______________________

______________________ ______________________

______________________ ______________________

I am a crab.
I eat clams, scallops,
and shrimp.

Here are some things that eat me:

I am a herring.
I eat phytoplankton
and zooplankton.

Here are some things that eat me:

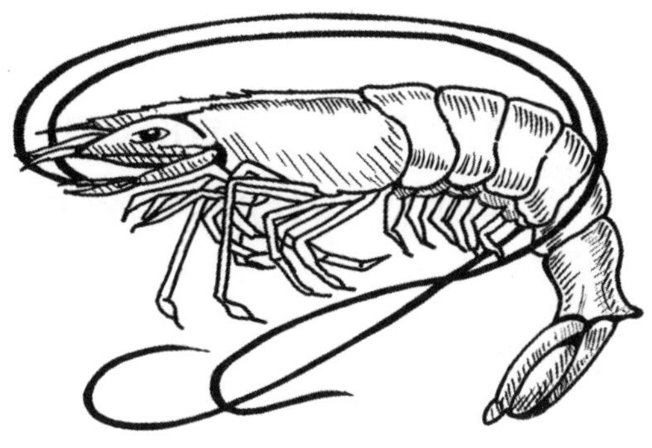

I am a shrimp.
I eat phytoplankton
and zooplankton.

Here are some things that eat me:

Here are some things that eat me:

I am a clam.
I eat phytoplankton.

Here are some things that eat me:

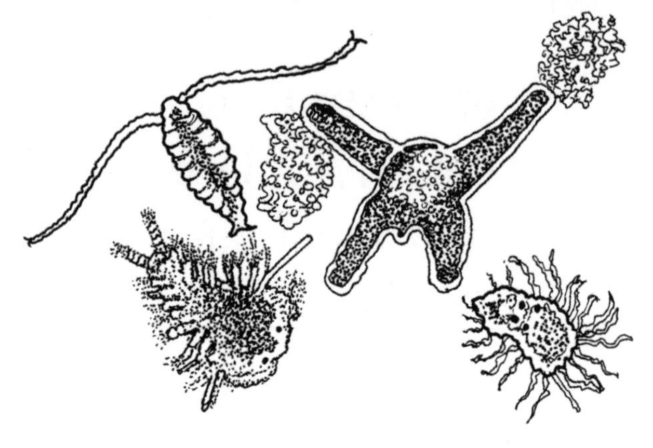

I am zooplankton.
I eat phytoplankton.

Here are some things that eat me:

I am a jellyfish.
I eat zooplankton, fish eggs, baby fish, and other small jellies.

Here are some things that eat me:

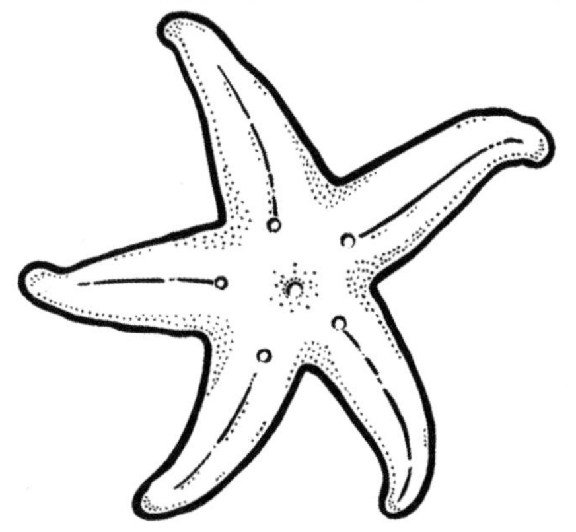

I am a sea star.
I eat phytoplankton, clams, scallops, and small fish.

Here are some things that eat me:

________________ ________________

________________ ________________

________________ ________________

________________ ________________

I am sea grass.
I need the sun to make my own food.

Here are some things that eat me:

________________ ________________

________________ ________________

________________ ________________

________________ ________________

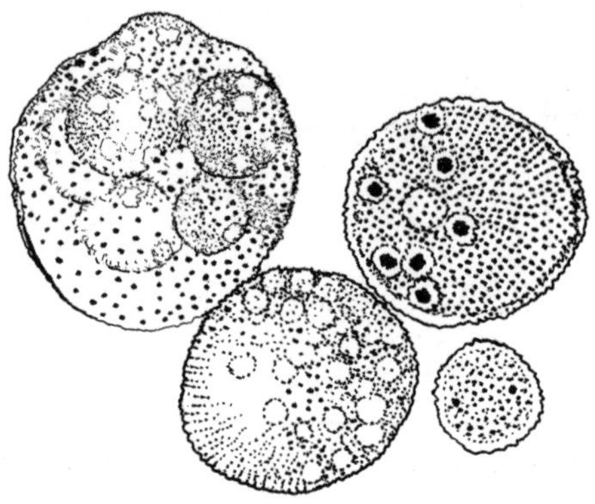

I am algae.
I need the sun to make my own food.

Here are some things that eat me:

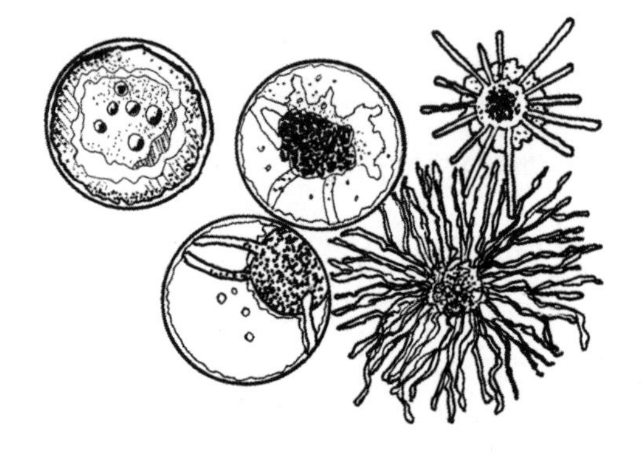

I am phytoplankton.
I need the sun to make my own food.

Here are some things that eat me:

Connecting Learning

Part One

1. Describe an ocean habitat.

2. What kinds of plants and animals live in that habitat?

3. Could any of the animals that we studied live in any other habitat? Explain your thinking.

4. What are some of the differences between the animals and what they eat?

5. What do the plants eat?

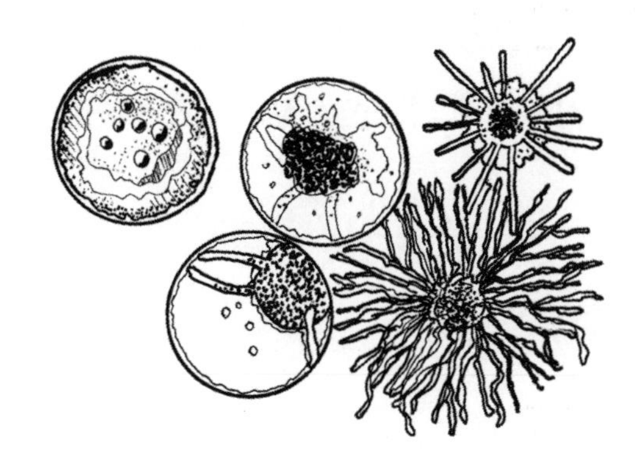

Connecting Learning

Part Two

1. How was our model like a real food web? How was it different?

2. What happened when plants and animals were taken out of the food web?

3. Which animals died first (without being removed by the teacher)?

4. Which animals survived the longest?

5. Which plants or animals had the biggest impact when they were removed? Why?

6. Which plants or animals had the least impact when they were removed? Why?

7. How do you think people fit in to our food web?

8. What are you wondering now?

Producers, Consumers, and Decomposers

1

8

All living things need food to survive. Green plants make their own food using energy from the sun. They are called **producers**. They produce food.

2

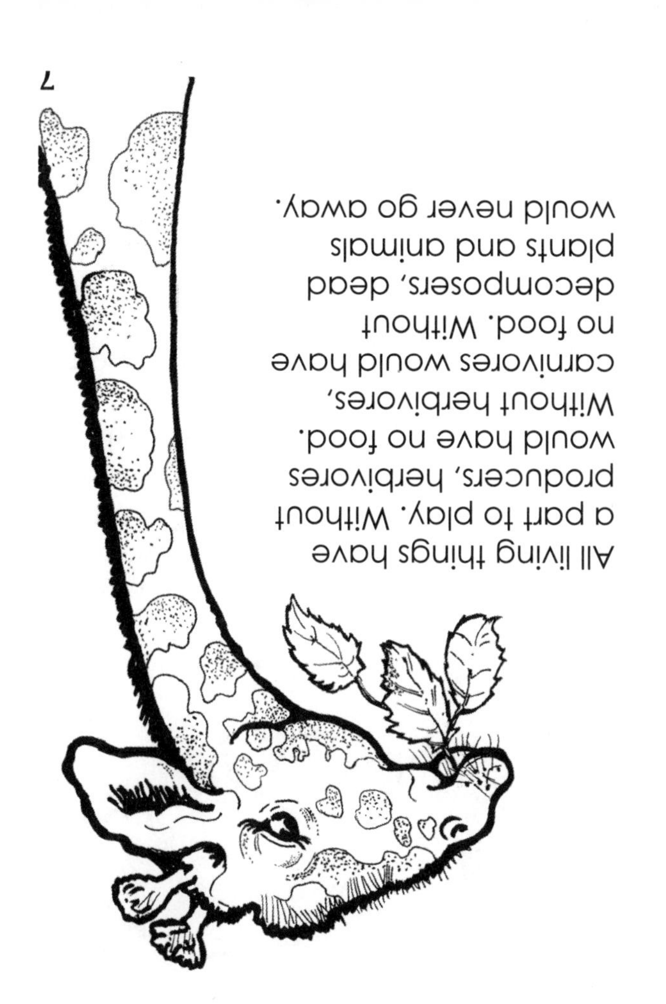

All living things have a part to play. Without producers, herbivores would have no food. Without herbivores, carnivores would have no food. Without decomposers, dead plants and animals would never go away.

7

Consumers eat plants and/or animals. They consume food. Some consumers only eat plants. They are called **herbivores.** Deer are herbivores. So are cows, aphids, giraffes, and horses.

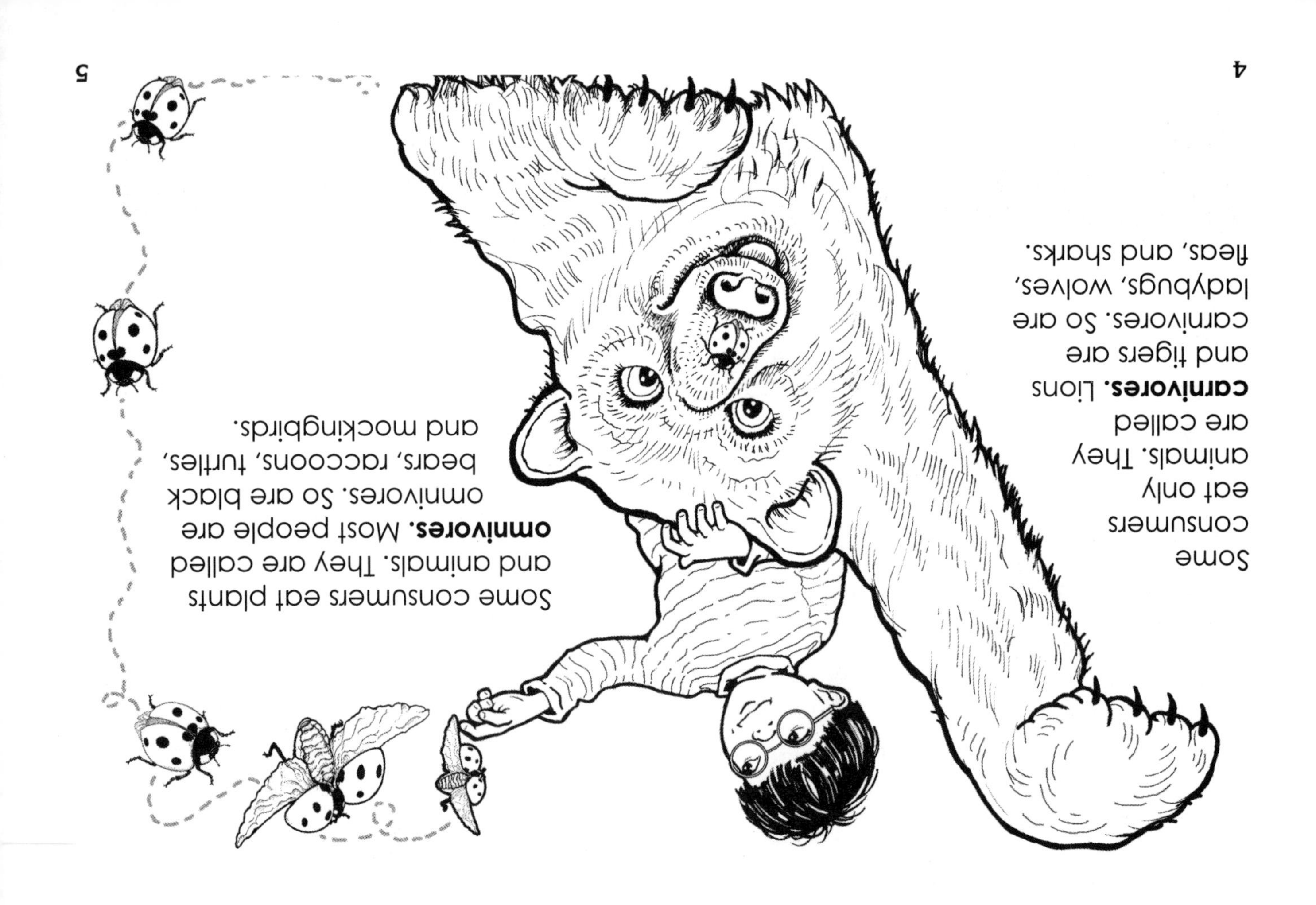

Some consumers eat only animals. They are called **carnivores.** Lions and tigers are carnivores. So are ladybugs, wolves, fleas, and sharks.

Some consumers eat plants and animals. They are called **omnivores.** Most people are omnivores. So are black bears, raccoons, turtles, and mockingbirds.

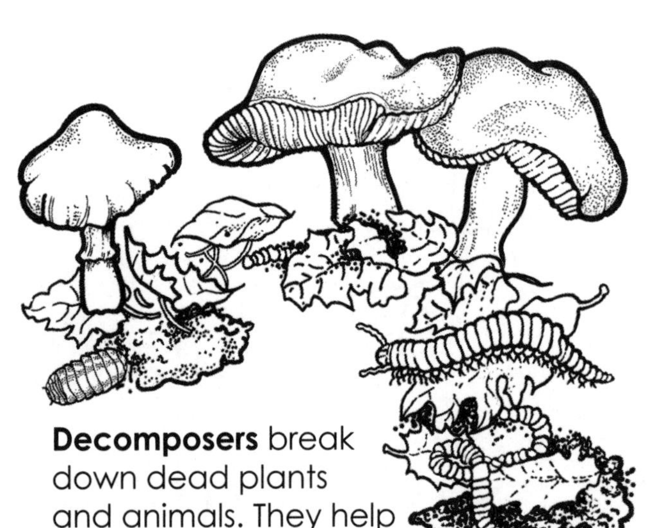

Decomposers break down dead plants and animals. They help return nutrients to the soil. There are many kinds of decomposers. Some are:

- Mushrooms
- Earthworms
- Bacteria
- Mold

Topic
Producers, consumers, and competition

Key Question
What happens when organisms at the same level of a food web compete with each other for food?

Learning Goals
Students will:
- understand the relationship between producers and consumers in food chains and food webs;
- recognize the distinctions between herbivores, carnivores, omnivores, and decomposers; and
- know that organisms may need to compete with each other for resources in an ecosystem.

Guiding Documents
Project 2061 Benchmarks
- *For any particular environment, some kinds of plants and animals survive well, some survive less well, and some cannot survive at all.*
- *Almost all kinds of animals' food can be traced back to plants.*
- *Some source of "energy" is needed for all organisms to stay alive and grow.*

NRC Standards
- *Organisms have basic needs. For example, animals need air, water, and food; plants require air, water, nutrients, and light. Organisms can survive only in environments in which their needs can be met. The world has many different environments, and distinct environments support the life of different types of organisms.*
- *All animals depend on plants. Some animals eat plants for food. Other animals eat animals that eat the plants.*

Science
Life science
 food chains and webs
  producers and consumers
  competition

Integrated Processes
Observing
Comparing and contrasting
Relating
Drawing conclusions
Applying

Materials
For the class:
 cardboard cut in 8 x 8 inch squares
  (see *Management 1)*

For each student group:
 game board and cards (see *Management 2)*
 tagboard or index cards, optional (see *Management 3)*
 small sticky note pad (see *Management 5)*
 scissors
 materials to make spinner (see *Management 1)*
 2-3 copies of the student page

Background Information
Animals can be broadly classified as consumers. This distinguishes them from green plants, which are producers. Students may have read or heard about other consumer categories. Consumers may be further classified according to their particular diet. For example, animals whose diet is primarily insects are called *insectivores*; those that eat mostly fruit are *frugivores*; those that eat seeds are *granivores*, and so forth. These categories are subgroups of carnivores and herbivores. Insectivores are a type of carnivore, and granivores and frugivores are both herbivores.

Most students have had plenty of experience with competition, although not likely for purposes of survival. In the natural world of an environment, consumers must compete for the food they need, not only with others of their own kind, but also with other species with a similar diet. Much of their time is spent in an ongoing quest to obtain enough food for themselves and their offspring to survive.

An animal population can only be as large as what can be supported by the available food and other resources in its environment. Competition is constant, and some animals will not get enough food to survive. If the supply of food in an environment is depleted for any reason, the competition becomes even more intense. Most consumers are *opportunists*—that is, while they may prefer certain types of food, they will settle for something else rather than go hungry. In so doing, they add to the number of animals already competing for that food.

Life in the natural world is "eat or be eaten." Very few animals do not have any predators. Even as an animal is eating or hunting for its next meal, it is at risk of becoming something else's meal instead.

Some competition scenarios:

- A deer population thriving on an abundance of food in a grassy mountain meadow is thrown into intense competition when much of the grass dies in an unseasonable dry spell. Only the healthiest, strongest animals will survive on whatever remaining grass or other food they can find. The weaker ones will starve or become sick and die.
- Several predators pursue a rabbit scampering across a field. Only one will get the meal. The winner will be the one that was fastest, or perhaps the most alert or most accurate.
- A large population of field mice supports a relatively large population of owls. A disease spreads through the mouse population, killing most of them. The owls must compete with other predators for voles, insects, or whatever they can find that they can eat. As a result, the owl population decreases.

Management

1. How to make a spinner:

   Eyelet spinner
   - Mount the game board on a sheet of cardboard.
   - Using a regular size paper clip, straighten the outer loop to make a pointer.
   - Poke a hole through the center of the game board.
   - Thread the loop of the paper clip onto the eyelet (available at fabric stores). Put it over the hole, broad side down against the board.
   - Put a paper fastener through the eyelet and the hole in the cardboard. Spread the fastener beneath the board to hold the spinner in place.

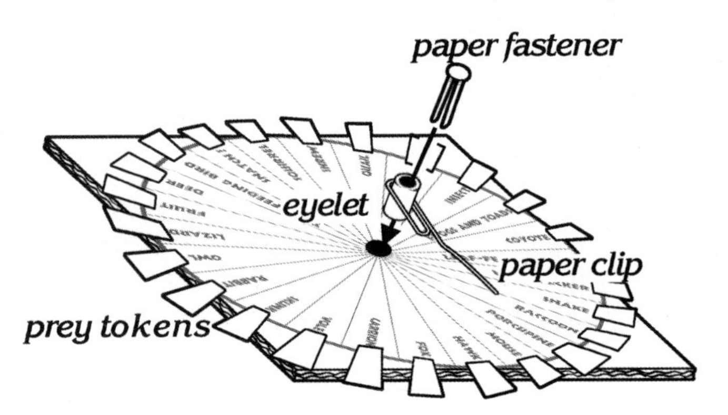

   Alternative—pushpin spinner
   - Glue a second piece of cardboard to the bottom of the game board to prevent the point of the pushpin from being a problem. Add a third piece of cardboard if necessary.
   - Poke a pushpin through the center of the game board. Pull it back out and put a bit of white glue on the point, then push it back in the hole and leave it for a few minutes.
   - Make a paper clip pointer as previously described.
   - Slide the inner loop of the pointer over the top of the pushpin.

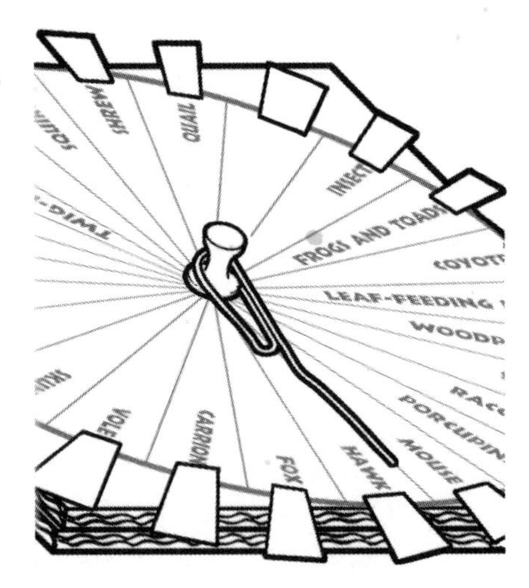

   - If the paper clip does not spin freely, an adult can broaden the inner loop somewhat by carefully inserting one scissors blade and turning it.
2. Corrugated cardboard cut from packing boxes is recommended for the backing of the game board.
3. The game cards can be made more durable by copying them on card stock. You may prefer to do this before doing the activity.
4. If you would like the game boards and cards to last for future use, laminate them before gluing the game board to the backing.
5. Students can cut their own "prey token" strips from the sticky notes. These are placed on the outer edge of each wedge to be collected by the first appropriate predator to land there.
6. The procedure is written with the "Carnivores Compete" game in mind, but any of the games may be used. For a more complete picture of how food chains function in an environment, at some point students should have the opportunity to play both games.
7. "Carnivores Compete" can be played by two to six people; "Omnivore Rivals" can be played by two to eight people.

Procedure

Part One

1. Tell the students to think about a meal they have eaten recently at home or in the school cafeteria. Who cooked the meal? [chef, cook, parent, sibling, student, etc.] Who ate the meal? [student, family, other students, etc.]
2. Have the students point out which people in this example made something and which ones used or ate it. Explain that in the natural world of the environment, there are those who *produce* and those who *consume*. However, in the environment, the *producers* produce the food within themselves, and the *consumers* consume the producer to get the food.
3. Review briefly what the students know about food chains and food webs.
4. Emphasize the names and roles of producers, consumers, carnivores, herbivores, omnivores, and decomposers, guiding students to understand what they do and how they are different from each other. Encourage the students to think of examples of each.

Part Two

1. If you have a monitor or helper system in your classroom, ask the students to name a few of the tasks. (If you don't have such a system, discuss one they've had in the past or might have.) List the tasks and define what each one does, e.g., *Messenger—takes messages to other classrooms.*
2. Ask what might happen if several people were assigned to the same job, but only one person could do it at a time. (Example: five are assigned to be messengers, but there is only one message to be taken.) Students are likely to focus on ways to be fair.
3. Add the following piece of information and discuss: Only the person who does the job gets to eat lunch. The others all have to go hungry.
4. If their responses still include taking turns, tell them that every single day the same person would get to do the job as well as eat lunch, and the others would go hungry. If competing to decide who gets the job has not come up, mention it as a possibility and discuss.
5. Ask the students what the word *competition* means to them. Direct the discussion to include sports, elections, contests, siblings, classmates, and whatever else may apply. In each case, discuss what factors are most likely to determine who wins.
6. Have the students think back to the monitor situation. Tell them the one who got the job and lunch did so only because he/she was the biggest or the strongest or the most alert or the most accurate. Explain that this is how it works in the natural world. Animals at each level of the food chain have a particular job to do in the environment, but they must compete with the others at that level to get the food and other resources they need.
7. Refer to the scenarios given in *Background Information* or create your own. Challenge the students to consider the consequences if there is not enough food to sustain all the consumers that need it. [get sick, starve, weaken and become easier prey, look for something else to eat, move to another location, etc.]
8. Tell the class that they will be playing a game to get a closer look at what it's like to be competing at one of the levels of the food chain. Distribute the games and materials and go over the rules. Have each group designate someone to keep track of what happens as the game progresses, using the student page.
9. After several rounds, compile the data from each group, discussing and looking for patterns and inconsistencies.

Connecting Learning

1. What is a producer? [an organism that makes its own food using energy from the sun] Give an example of a producer. [grass, tree, bush, etc.]
2. What is a consumer? [an organism that eats plants and/or animals] Give an example of a consumer. [people, dogs, cats, cows, etc.]
3. Which plants or animals were left at the end or were the last to go? Were the same ones almost always left or last in every game played? How could you explain this? [needed by fewer predators, plus luck of spin]
4. Which plants or animals were "consumed" early in the game? Was this consistent in multiple games? How might this be explained?
5. At the end of the game, what were the chances of survival for the animal you represented? Explain.
6. If you could remove one animal from the competition to give your animal a better chance of survival, which animal would it be? Why? (be specific)
7. What changes would you make to the game board to give your animal a better chance to survive? Explain your thinking.
8. What happens when organisms at the same level of a food web compete with each other for food?
9. What are you wondering now?

Extensions

1. Keeping the same groups, have the students play each game and compare the results.
2. Duplicate extra cards and play the game with all players representing the same animal. Compare the results.
3. Design game boards for other environments—playground, backyard, desert, pond, tide pool, etc.

Curriculum Correlation

Lauber, Patricia. *Who Eats What?* Harper Collins. New York. 1995.
Clear explanation and examples of representative food chains.

Powell, Consie. *A Bold Carnivore: An Alphabet of Predators.* Raven Productions, Inc. Ely, MN. 2007.
A North American carnivore for each letter of the alphabet.

Internet Connections

Carnivores
http://www.nhptv.org/natureworks/nwep10a.htm

Decomposers
http://www.nhptv.org/natureworks/nwep11b.htm

Herbivores
http://www.nhptv.org/natureworks/nwep9a.htm

Omnivores
http://www.nhptv.org/natureworks/nwep10b.htm

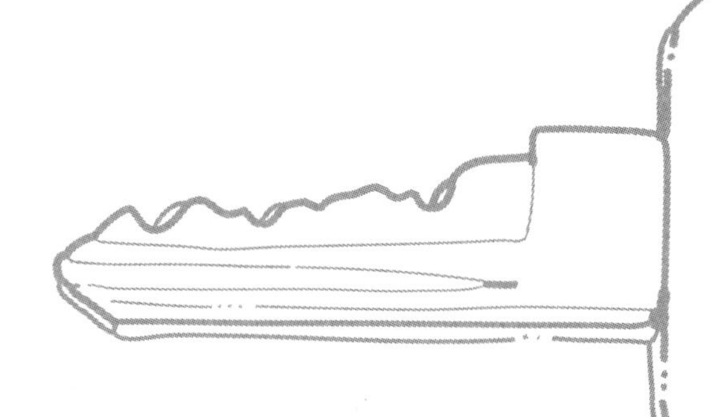

Key Question

What happens when organisms at the same level of a food web compete with each other for food?

Learning Goals

Students will:

- understand the relationship between producers and consumers in food chains and food webs;
- recognize the distinctions between herbivores, carnivores, omnivores, and decomposers; and
- know that organisms may need to compete with each other for resources in an ecosystem.

Carnivore Competition

for 2 to 6 players

Rules

1. Stick *Prey Token* strips around the edge of the game board, one strip for each wedge.
2. Shuffle the *Predator Cards* face down and deal one to each player, or let each player draw one from the stack.
3. The player with the first *Predator Card* in alphabetical order goes first, and play continues around the circle to the left.
4. Each player spins in turn. When the spinner stops on a prey wedge, the player checks his or her *Predator Card*. If that particular prey is listed as being one the predator will eat, the player removes the *Prey Token* from the board and keeps it. That particular prey is considered to be consumed and is no longer available to any predators.
5. If a player lands on a prey wedge not listed on his or her *Predator Card*, the turn moves on to the next player.
6. If a player lands on a prey wedge from which the *Prey Token* has already been taken, the turn moves on to the next player.
7. Any time a player lands on the *Grab* wedge, he or she may take a token from the player on his/her right. If that player doesn't have a token yet, he/she will have to hand over the next one obtained.
8. The game is over when either all the *Prey Tokens* have been removed (i.e., captured and consumed) or three consecutive rounds are played without anyone being able to pick up a *Prey Token*. If a player runs out of available prey before the end of the game, he or she will need to wait until the game has run its course.

The winner is the predator with the most *Prey Tokens.*

Interpretation

The player with the most tokens was the most successful at competing for food. A player with a comparatively large number of tokens probably has enough food for the time being.

A player with only a few tokens is in danger of starvation.

A player left with no tokens was unable to find the food needed to survive.

Optional

Carnivores Compete: At the teacher's discretion and to make the game a bit more competitive, special rules may apply to the Owl, Bobcat, and Mountain Lion wedges. These high-level predators are preyed upon by certain other high-level predators. So, the first time

- the bobcat player lands on the hawk or owl wedge
- the mountain lion lands on the coyote or fox wedge, or
- the owl lands on the hawk wedge,

the predator takes the *Prey Token* as usual. After that, any time the bobcat, mountain lion, or owl lands on the empty wedge of the hawk, owl, coyote or fox wedge—if it is one of its prey—the player representing the prey must give one of his or her *Prey Tokens* to the predator. If the prey is not represented by a player, the wedge is treated like any other.

Omnivore Rivals

for 2 to 8 players

Rules

1. Follow the rules for *Carnivore Competition* using the omnivore game board and omnivore cards.

Optional

Similar special rules may apply to the wedges. Some of these omnivore predators are preyed upon by certain other omnivore predators. So, the first time the opossum, raccoon, or skunk player lands on the vole wedge, or the opossum or skunk lands on the hawk wedge, the predator takes the *Prey Token* as usual. After that, any time the opossum, raccoon, or skunk lands on the empty wedge of the shrew or vole—if it is one of its prey—the player representing that prey must give one of his or her *Prey Tokens* to the predator. If the prey is not represented by a player, the wedge is treated like any other.

Note for Omnivore Rivals

- A bird eaten by an animal in this group is likely to be an egg or young bird stolen from a nest.

General Notes

- In this context, *herb* means *herbaceous plant*, which is any seed-producing flowering plant that does not have woody stems. It may be either an annual plant or one that produces new stems and leaves each growing season.
- A *leaf-feeding* bird is one so small that it can actually flit onto a leaf to grab an insect.
- A *twig-feeding* bird is generally larger and finds its food on twigs that it can sit on.
- There are some twig-feeders that are small, but they find their food on twigs rather than leaves. Many birds also catch insects while on the wing, but for the purpose of food chains, these two categories give a fairly good idea of the size of the prey.
- Sometimes a predator catches its prey only to have the prey taken away from it by another predator. The *Grab* wedge represents this occurrence.
- Some predators also prey upon other predators competing for food at the same level. Thus some animals appear on both *Predator Cards* and *Prey Wedges*. The optional rules give more emphasis to this situation.

Game Board
Carnivore Competition

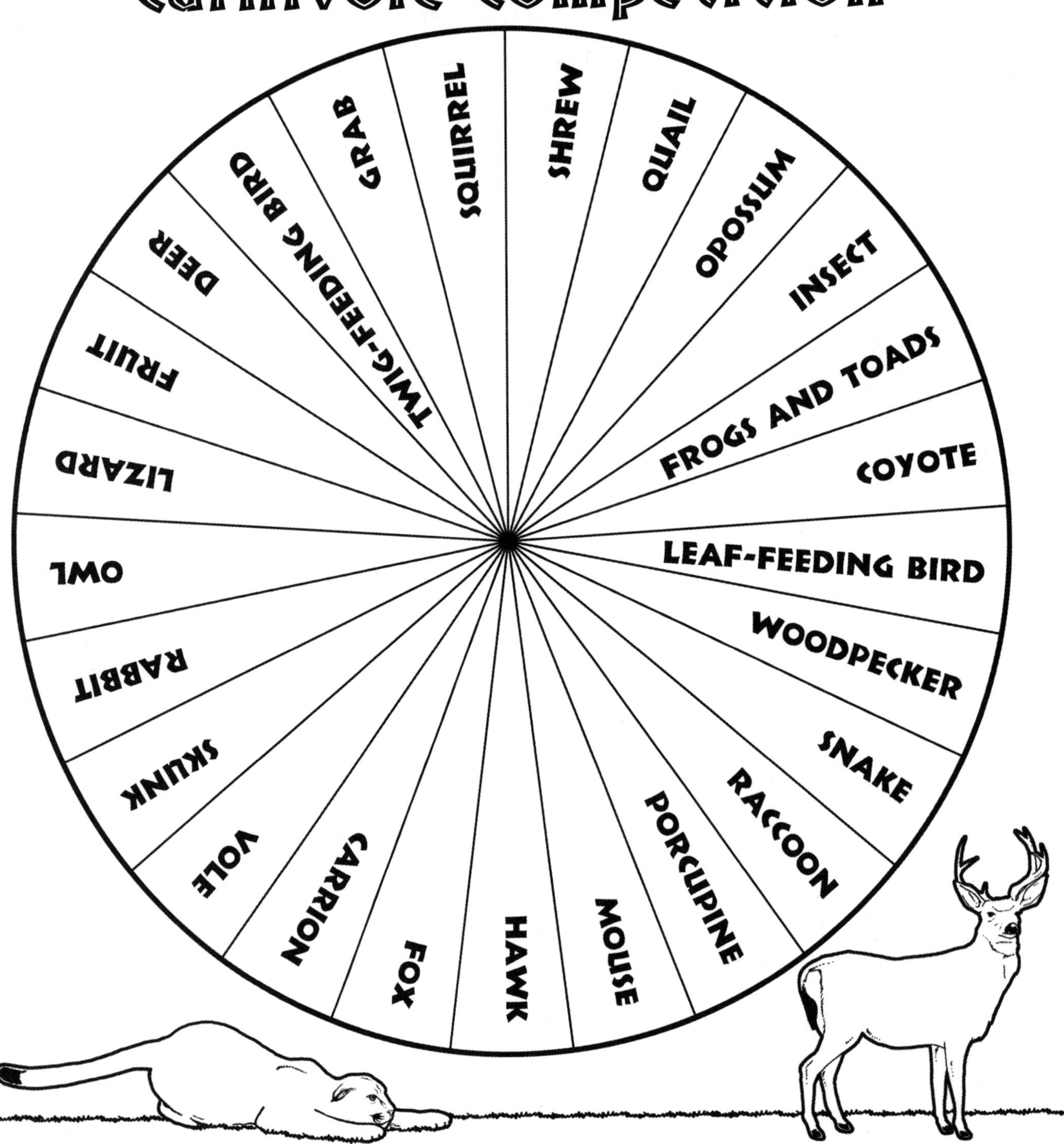

Game Board
Omnivore Rivals

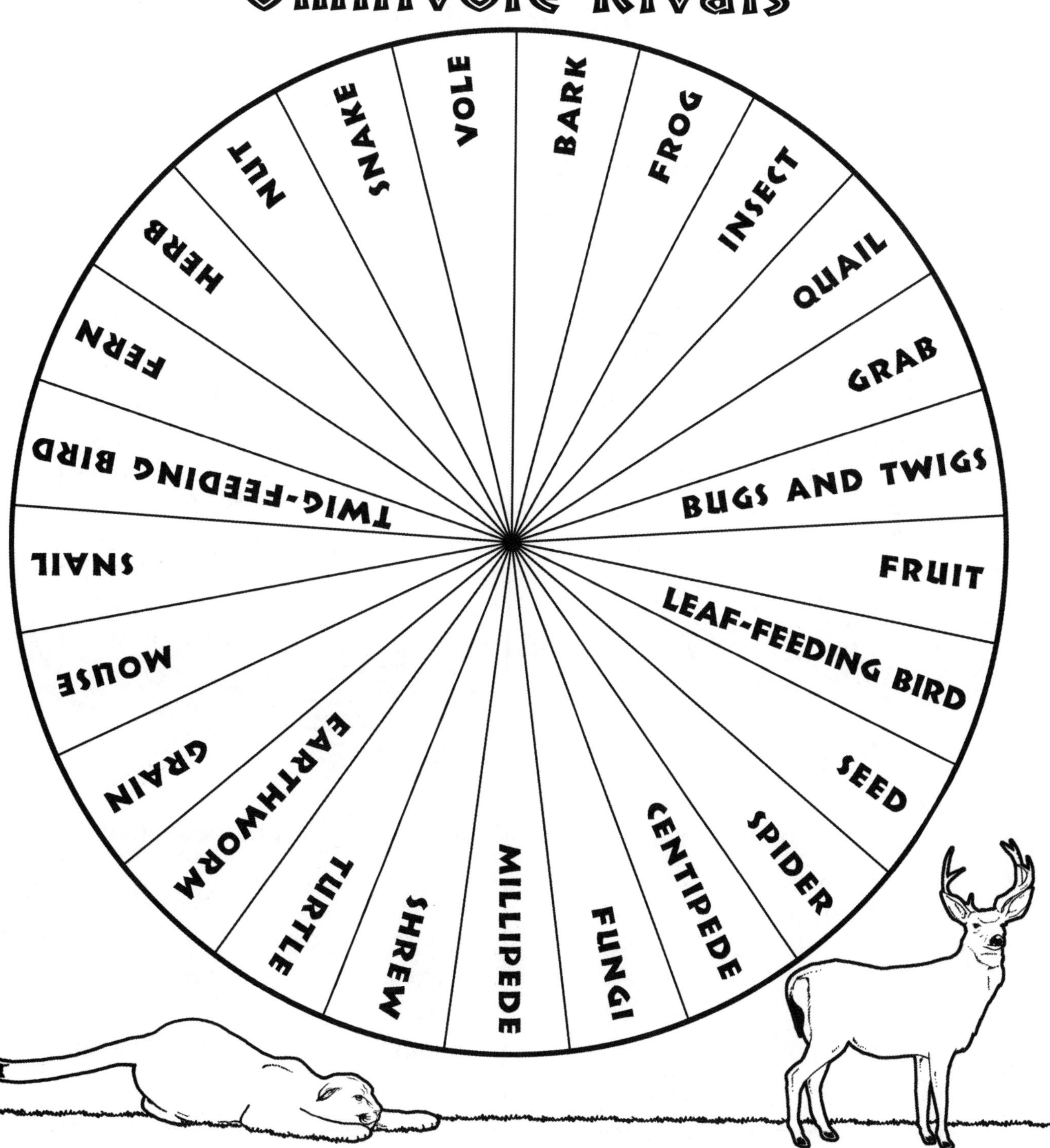

Carnivore Competition

These high-level predators are carnivores (meat eaters). The coyote and fox are omnivorous in that they will also eat fruit. These predators are hunters and rovers, covering a sizeable territory in their quest for food. These six animals are only a few of those who are found at the top of the food chain in a particular habitat. Not all high-level predators are mammals. For example, a shark holds a comparable position in an ocean environment.

Omnivore Rivals

There are many other omnivores, both vertebrates and invertebrates, feeding at this level in any given habitat. The list of predators might include many reptiles, birds, amphibians, mammals, and even insects and other invertebrates. Omnivorous fish and other animals fill this niche in aquatic environments.

For purposes of simplicity, this representation of omnivores consists entirely of vertebrate animals.

Carnivore Predator Cards

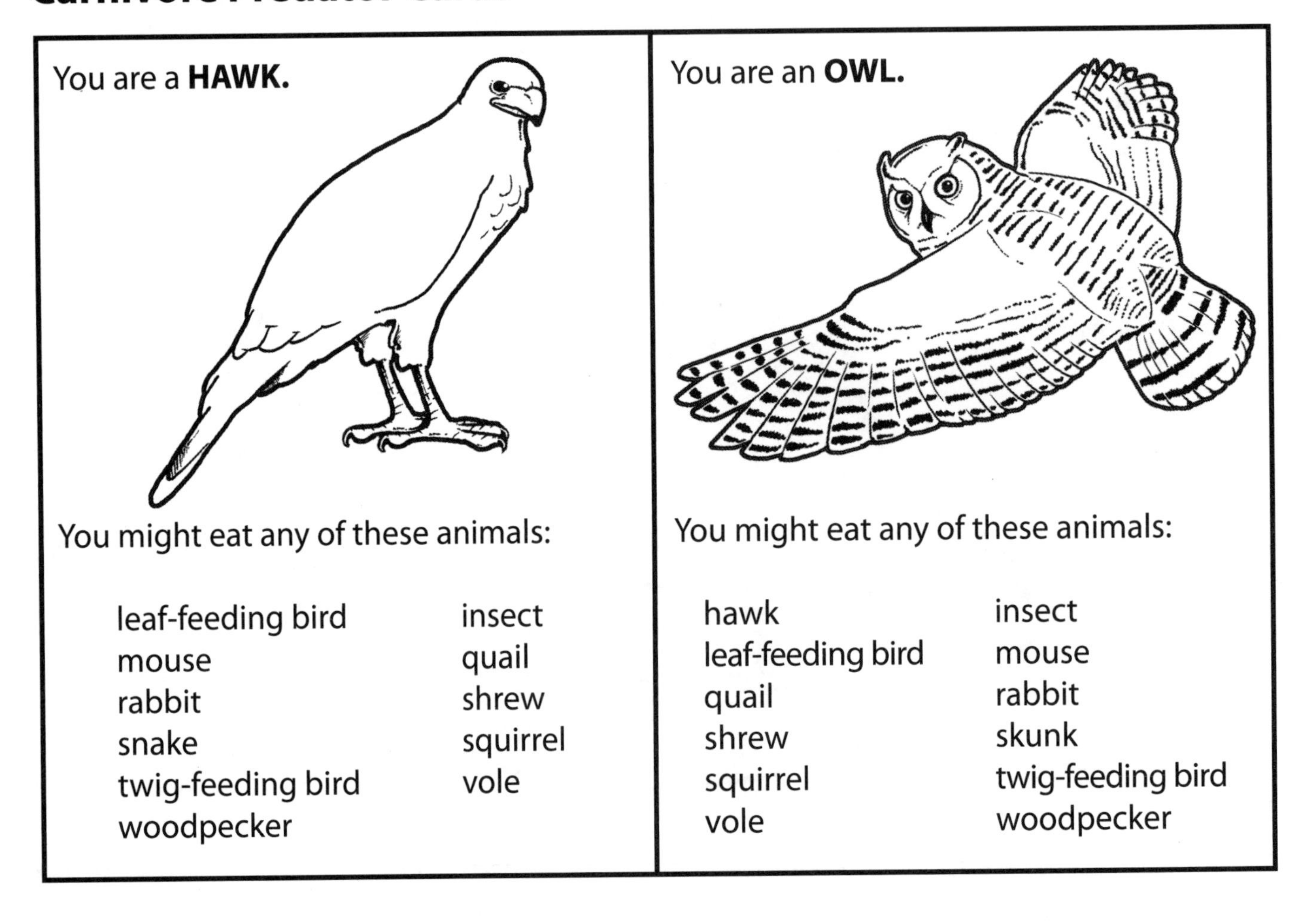

Carnivore Predator Cards

You are a **BOBCAT.**

You might eat any of these animals:

- hawk
- mouse
- owl
- quail
- raccoon
- skunk
- woodpecker
- insect
- opossum
- porcupine
- rabbit
- shrew
- squirrel
- twig-feeding bird

You are a **MOUNTAIN LION.**

You might eat any of these animals:

- coyote
- fox
- frogs and toads
- mouse
- rabbit
- shrew
- vole
- deer
- insect
- lizard
- opossum
- raccoon
- squirrel

You are a **FOX.**

You might eat any of these animals:

- deer
- frogs and toads
- opossum
- quail
- shrew
- snake
- insects
- mouse
- porcupine
- rabbit
- skunk
- squirrel

You might also eat:

- fruit

You are a **COYOTE.**

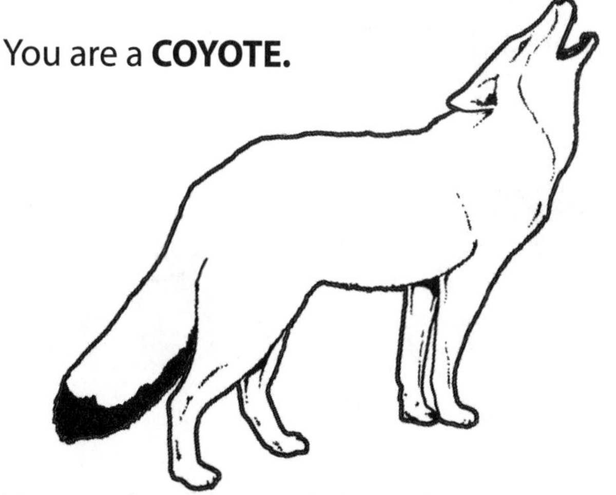

You might eat any of these animals:

- lizard
- rabbit
- mouse
- vole
- carrion (dead animals)

You might also eat:

- fruit

Omnivore Predator Cards

You are a **MOUSE.**

You might eat any of these things:

centipede	fruit
grain	herb
insect	millipede
nut	seed
snail	spider

You are a **SHREW.**

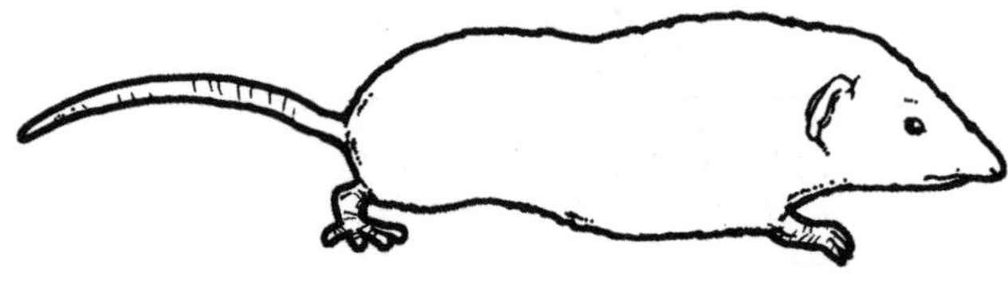

You might eat any of these things:

centipede	earthworm
frog	grain
insect	millipede
mouse	snail
spider	

You are a **RACCOON.**

You might eat any of these things:

earthworm	fruit
frog	leaf-feeding bird
insect	mouse
nut	seed
shrew	snail
vole	turtle
twig-feeding bird	

You are an **OPOSSUM.**

You might eat any of these things:

earthworm	fruit
insect	mouse
leaf-feeding bird	nut
quail	seed
shrew	snail
snake	twig-feeding bird
vole	

Omnivore Predator Cards

You are a **CROW.**

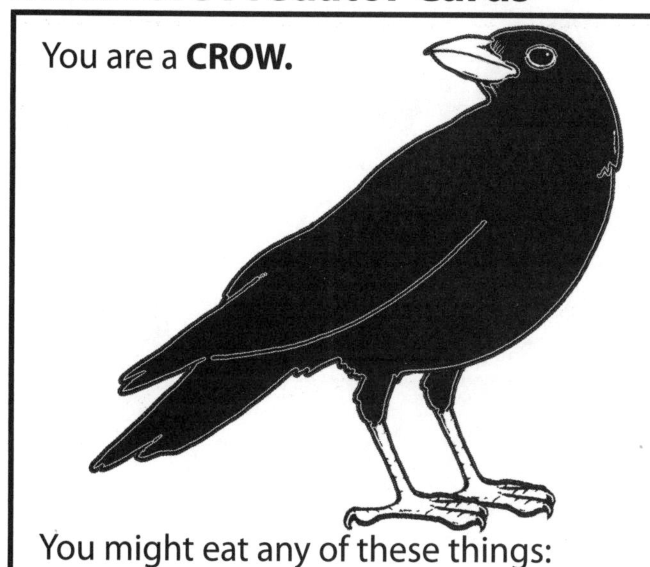

You might eat any of these things:

centipede	earthworm
frog	fruit
grain	insect
millipede	nut
seed	snail
spider	

You are a **SQUIRREL.**

You might eat any of these things:

bark	buds and twigs
fern	fruit
fungi	grain
herb	insect
leaf-feeding bird	nut
quail	seed
twig-feeding bird	

You are a **SKUNK.**

You might eat any of these things:

centipede	frog
fruit	fungi
insect	mouse
quail	snail
snake	spider
turtle	vole

You are a **VOLE.**

You might eat any of these things:

buds and twigs	centipede
earthworm	grain
herb	insect
millipede	seed
snail	spider

Game played: ______________________

1. Which were the first five items to be consumed?

2. Which items were left on the board at the end?

3. Which items seemed to be most in demand?

4. List the animals and the number of prey caught by each. Tell whether you think each would survive, probably would survive, or might not survive.

CONNECTING
LEARNING

Connecting Learning

1. What is a producer? Give an example of a producer.

2. What is a consumer? Give an example of a consumer.

3. Which plants or animals were left at the end or were the last to go? Were the same ones almost always left or last in every game played? How could you explain this?

4. Which plants or animals were "consumed" early in the game? Was this consistent in multiple games? How might this be explained?

5. At the end of the game, what were the chances of survival for the animal you represented? Explain.

CONNECTING LEARNING

Connecting Learning

6. If you could remove one animal from the competition to give your animal a better chance of survival, which animal would it be? Why?

7. What changes would you make to the game board to give your animal a better chance to survive? Explain your thinking.

8. What happens when organisms at the same level of a food web compete with each other for food?

9. What are you wondering now?

Topic
Interdependence

Key Question
How are the animals of the rain forest dependent on the trees?

Learning Goals
Students will:
- recognize that the plants of the rain forest provide food and shelter for all the creatures that live there, and
- identify the food chain that makes each animal dependent on the plants within the habitat for food.

Guiding Documents
Project 2061 Benchmarks
- *A great variety of kinds of living things can be sorted into groups in many ways using various features to decide which things belong to which group.*
- *Features used for grouping depend on the purpose of the grouping.*
- *Changes in an organism's habitat are sometimes beneficial to it and sometimes harmful.*
- *Almost all kinds of animals' food can be traced back to plants.*

NRC Standards
- *Populations of organisms can be categorized by the function they serve in an ecosystem. Plants and some micro-organisms are producers—they make their own food. All animals, including humans, are consumers, which obtain food by eating other organisms. Decomposers, primarily bacteria and fungi, are consumers that use waste materials and dead organisms for food. Food webs identify the relationships among producers, consumers, and decomposers in an ecosystem.*
- *For ecosystems, the major source of energy is sunlight. Energy entering ecosystems as sunlight is transferred by producers into chemical energy through photosynthesis. That energy then passes from organism to organism in food webs.*

Science
Life science
  interdependence
  food chains

Integrated Processes
Observing
Classifying
Comparing and contrasting
Generalizing

Materials
Student pages
Colored pencils
Scissors
Glue sticks
The Great Kapok Tree (see *Management 3)*

Background Information
This investigation is based on the book *The Great Kapok Tree: A Tale of the Amazon Rain Forest* by Lynne Cherry (Harcourt Brace Jovanovich. New York. 1990). Reading the story to the students sets the stage for further study.

All life in a habitat is formed around the plants that thrive with the sunlight and water available. In a tropical rain forest, the climax plants are the tallest emergent trees that rise high above the jungle floor. In the Amazon rain forest, the kapok tree is one of these plants. It often rises 100 feet above the ground with its crown rising above the surrounding trees. Below its branches is the canopy of the rain forest, formed by the dense top growth of smaller trees. This canopy thrives with animal and plant life of every form. The canopy blocks sunlight and shelters the plants and animals below from the intense sun of the tropics. Lush shrubs grow densely below the canopy in the humid, sheltered environment. This understory forms a short forest of its own below the canopy. The floor of the rain forest supports small plants and flowers that prefer the shady environment.

All life in the forest is dependent on the plants symbolized by the most dramatic specimen, the kapok tree. Remove the major plant life and the whole habitat collapses. The plants provide the shelter for all animal life in the forest as well as returning oxygen to the atmosphere through photosynthesis. Photosynthesis allows the plants to convert sunlight, water, carbon dioxide, and nutrients into the energy required by the rest of the forest life.

Many animals in the rain forest eat the plants directly to get energy. These herbivores are the primary consumers in the forest. Other animals get their food indirectly from the plants by eating the herbivores. These omnivores and carnivores are the secondary consumers in the habitat. If the plants are destroyed, the whole chain of dependence is destroyed.

The plants support the animals in the forest but are also dependent on other forms of life. The decomposers, mostly bacteria and fungi, convert plant and animal waste products into nutrients for plants. Many plants depend on animals for pollination and seed dispersion. Dependence in the forest is a two way street with each organism giving to and receiving from other organisms.

Management

1. This activity is an excellent culminating investigation in habitats or interdependence.
2. This investigation assumes students are familiar with interdependence and the concepts of producer and consumer, as well as herbivore, omnivore, and carnivore. As a culminating investigation, it provides an excellent opportunity to assess student understanding.
3. It works best to have multiple copies of *The Great Kapok Tree.*

Procedure

Part One

1. Ask the *Key Question* and state the *Learning Goals.*
2. Read or have students read the book *The Great Kapok Tree: A Tale of the Amazon Rain Forest* by Lynne Cherry. Have students discuss why the tree is so important to the animals. Be sure to have students study the illustrations on the inside covers. They are very informative as to where rain forests exist, and where rain forests once existed.
3. Distribute the chart of animal information and the illustrated stamps of the animals. Using the information, have the students identify and appropriately color each animal. Direct them to cut out one set of animal stamps to glue on the chart.
4. Have the students cut out another set of animal stamps and sort them on the *Food* page by type of consumer and glue them in the appropriate position.
5. Have students draw arrows from each stamp to the type of food it eats. The herbivores—porcupine, butterfly, and sloth—will have arrows drawn to the tree's crown. The omnivores—monkey and toucan—will have arrows drawn to the tree and butterfly (insect). The carnivores will have arrows drawn only to other animals.
6. Have the students cut out the remaining set of stamps and sort them on the *Shelter* page by the animal's location in the forest and glue the stamps in the appropriate position.
7. Read the book a second time, a page at a time, and have students record arrows to show the different dependencies (shelter, pollination, soil maintenance, food, oxygen) described on each page.

Part Two

1. A much more exciting investigation is made by having each student research a specific rain forest animal. There are a number of animals not mentioned in the book as well as many species of butterflies, birds, reptiles, monkeys, and tree frogs. Have students find information and use their literary skills to write a report about the animal's size, appearance, food sources, shelter location, and other interesting facts. Using their math measurement skills, have them build a life-size model of the animal. (Students with butterflies or tree frogs might be encouraged to make several models.)
2. A rain forest can be built in the classroom. An emergent tree can be made in the middle of the classroom by wrapping a carpet roll with brown butcher paper. The roll should be secured by placing it in a bucket filled with sand or rocks and buttressed with a group of student desks. The top of the roll can be secured to the ceiling with green paper in the shape of large leaves for the crown. The canopy trees can be made in the corner of the room out of crumpled butcher paper and broad leaves attached to the walls. The understory can be made by attaching leaves and greenery to the desks around the room.
3. As each student presents his or her animal to the class, it can be hung in the appropriate level of the rain forest in the room. The food chain can be shown by stringing yarn between the appropriate animals. When all the animals have been placed and the food chain strung, the students can summarize what they learned by completing the record pages and having a closing discussion about interdependencies.

Connecting Learning

1. How is each animal dependent on plants for shelter?
2. What producers are in the story?
3. What consumers are in the story? Which consumers are herbivores? ...carnivores? ...omnivores?
4. Why will the jaguar (anteater) suffer if the rain forest is cut down? [no food, no primary consumers to eat since plants are gone]

5. How would the trees suffer if all the animals were killed? [no pollination, no waste nutrients]
6. How are humans, the man and boy, dependent on the tree? [oxygen, material, food, medicine, beauty]
7. How are the animals in your neighborhood dependent on the plants for survival?
8. How can humans get the wood they need without destroying the rain forest?
9. What are you wondering now?

Curriculum Correlation

Literature

Cherry, Lynne. *The Great Kapok Tree: A Tale of the Amazon Rain Forest.* Harcourt Brace Jovanovich. New York. 1990.

Gibbons, Gail. *Nature's Green Umbrella.* HarperCollins Juvenile Books. Toronto. 1997.

Pratt, Kristin Joy. *A Walk in the Rainforest.* Dawn Publications. Nevada City, CA. 1992.

Silver, Donald M. *One Small Square: Tropical Rain Forest.* Learning Triangle Press. New York. 1999.

Stille, Darlene R. *Tropical Rain Forests.* Children's Press. Chicago. 2000.

Video

National Geographic Society. *National Geographic's Really Wild Animals: Totally Tropical Rain Forest.* National Geographic. Washington, DC. 1994.

The Kapok Tree

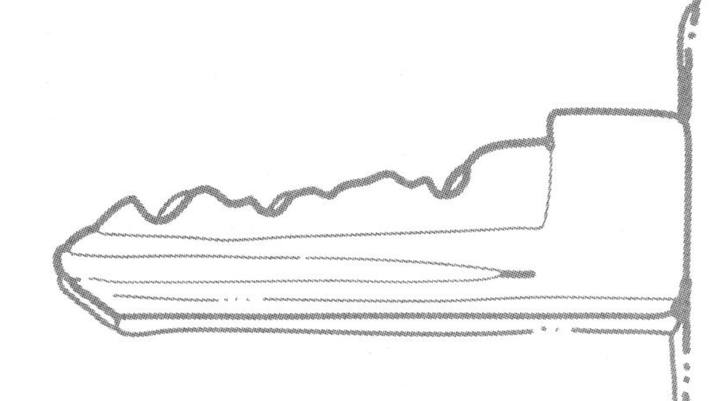

Key Question

How are the animals of the rain forest dependent on the trees in the jungle?

Learning Goals

Students will:

- recognize that the plants of the rain forest provide food and shelter for all the creatures that live there, and
- identify the food chain that makes each animal dependent on the plants within the habitat for food.

The Kapok Tree Stamps

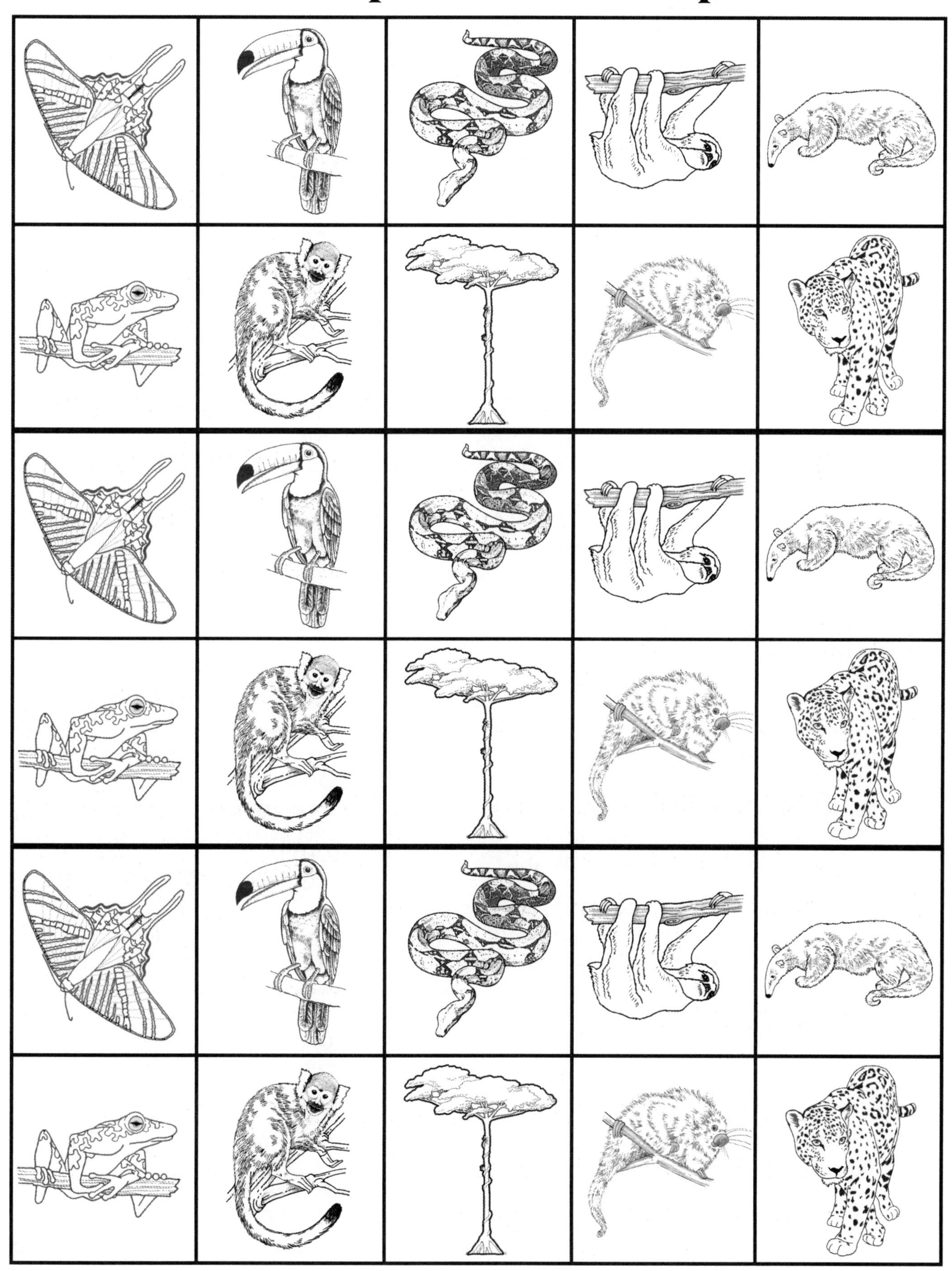

Rain Forest Animal Chart

Animal	Size / Appearance	Food / Active / Location
Jaguar (mammal)	**Size** 1.5-2 m long, 0.5-1 m tail; **Appearance** cat, brown-yellow fur with black spots	**Food** sloth, reptiles, birds/eggs; **Active** nocturnal; **Location** floor
Toucan (bird)	**Size** 50 cm long, beak is 1/3 of length; **Appearance** huge beak, yellow chest with black body	**Food** fruit, insects; **Active** diurnal; **Location** emergent, canopy
Boa Constrictor (reptile)	**Size** 60 cm long baby, 2-3.5 m adult, 6m maximum; **Appearance** broken pattern of cream, tan, brown, and black	**Food** birds, monkeys, lizards; **Active** nocturnal; **Location** canopy
Sloth (mammal)	**Size** 60 cm, about the size of a cat; **Appearance** thick brown fur, slightly green	**Food** leaves, shoots, flowers; **Active** nocturnal; **Location** canopy

Animal	Size / Appearance	Food / Active / Location
Animal **Butterfly (insect)**	**Size** 5-20 cm **Appearance** brightly colored wings	**Food** plant sap and nectar, leaves (caterpillar) **Active** diurnal **Location** understory
Animal **Monkey (mammal)**	**Size** 3-10 cm long **Appearance** furry with long, grasping tail	**Food** leaves, fruit, nuts, bird eggs, insects **Active** diurnal **Location** canopy
Animal **Tree frog (amphibian)**	**Size** 5-8 cm **Appearance** colorful (green, blue, red, orange), long rear legs	**Food** insects, small invertebrates **Active** nocturnal **Location** canopy
Animal **Tree Porcupine (mammal)**	**Size** head and body 50 cm, tail 45 cm **Appearance** long tail, body covered with short spines	**Food** leaves, stems, fruits **Active** nocturnal **Location** canopy
Animal **Anteater (mammal)**	**Size** Giant 3 m Lesser 1.5 m Pygmy 40 cm **Appearance** long snout and tongue, coarse fur	**Food** ants and termites (insects) **Active** nocturnal **Location** floor, understory, canopy

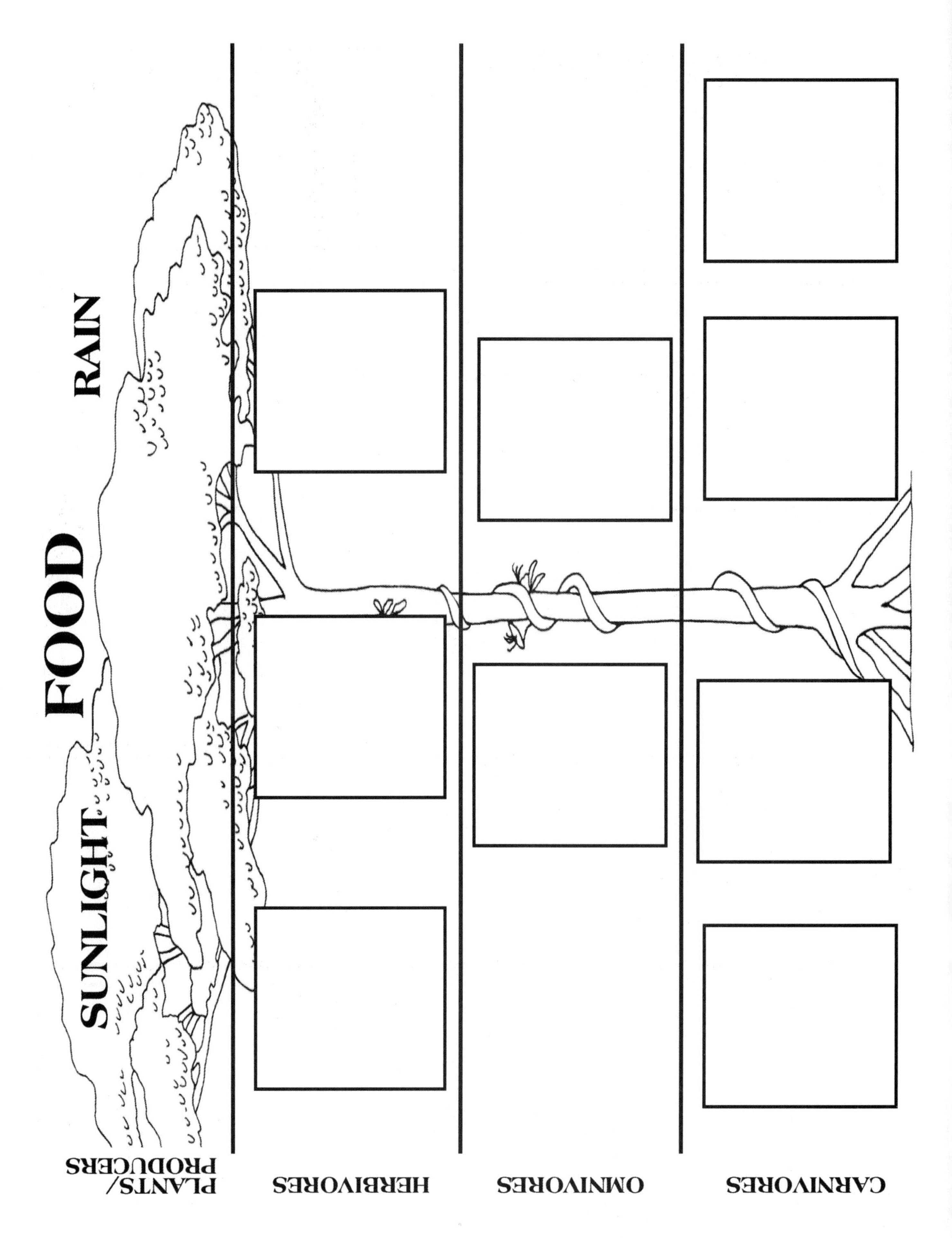

RAIN
FOOD
SUNLIGHT
PLANTS/
PRODUCERS
HERBIVORES
OMNIVORES
CARNIVORES

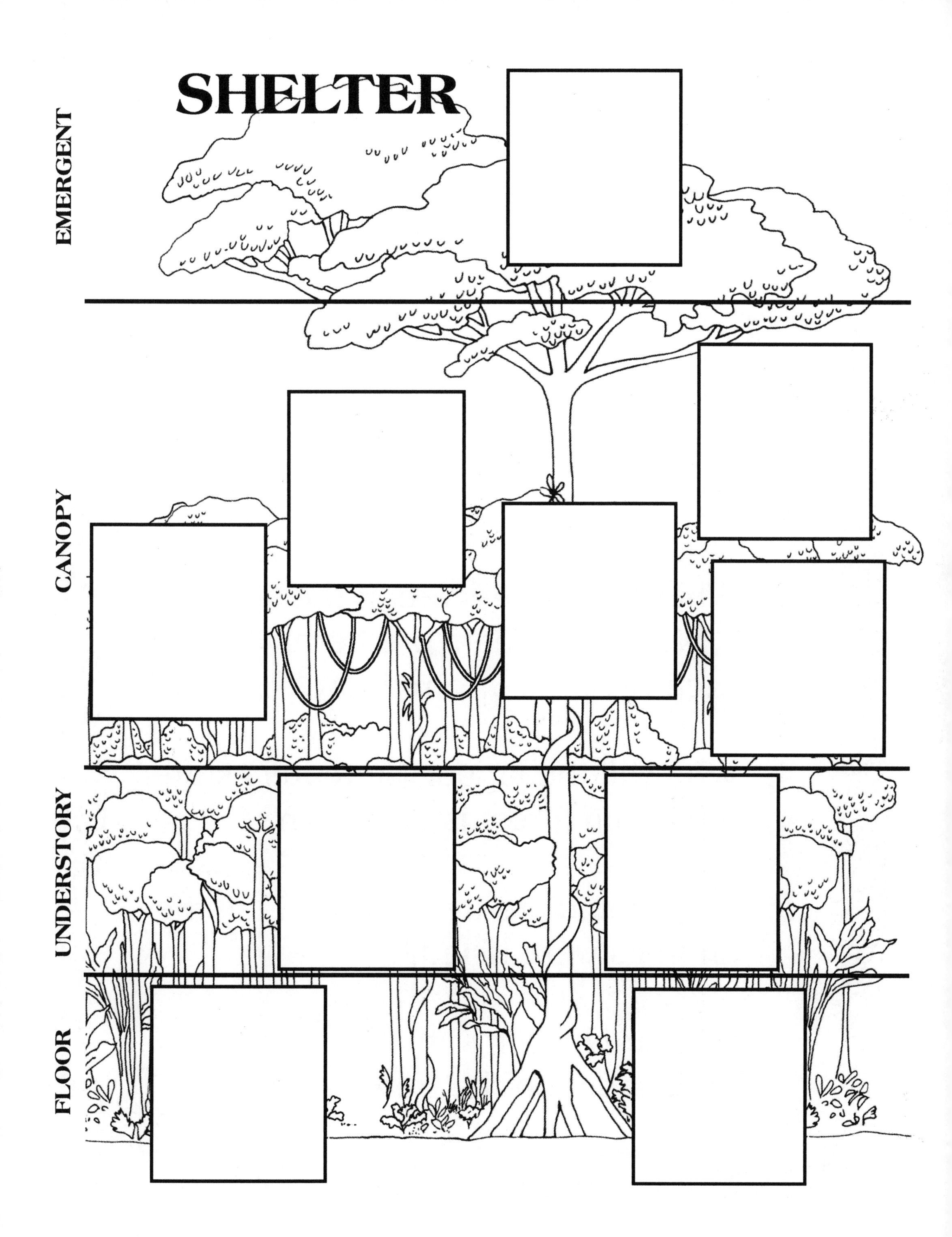
SHELTER
EMERGENT
CANOPY
UNDERSTORY
FLOOR

The Kapok Tree

Connecting Learning

1. How is each animal dependent on plants for shelter?

2. What producers are in the story?

3. What consumers are in the story? Which consumers are herbivores? ...carnivores? ...omnivores?

4. Why will the jaguar (anteater) suffer if the rain forest is cut down?

5. How would the trees suffer if all the animals were killed?

6. How are humans, the man and boy, dependent on the tree?

The Kapok Tree

Connecting Learning

7. How are the animals in your neighborhood dependent on the plants for survival?

8. How can humans get the wood they need without destroying the rainforest?

9. What are you wondering now?

What's the Net Worth?

Topic
Interdependence

Key Question
Using the Brazil Nut tree as an example, how would you explain the importance of protecting the trees in the rain forest?

Learning Goal
Students will model the complex biodiversity, energy levels, and interdependence in a rain forest by role-playing the plants and animals in a rain forest food web.

Guiding Document
Project 2061 Benchmarks

- *Human activities, such as reducing the amount of forest cover, increasing the amount and variety of chemicals released into the atmosphere, and intensive farming, have changed the earth's land, oceans, and atmosphere. Some of these changes have decreased the capacity of the environment to support some life forms.*
- *One of the most general distinctions among organisms is between plants, which use sunlight to make their own food, and animals which consume energy-rich foods. Some kinds of organisms, many of them microscopic, cannot be neatly classified as either plants or animals.*
- *All organisms, including the human species, are part of and depend on two main interconnected global food webs. One includes microscopic ocean plants, the animals that feed on them, and finally the animals that feed on those animals. The other web includes land plants, the animals that feed on them, and so forth. The cycles continue indefinitely because organisms decompose after death to return food material to the environment.*
- *Two types of organisms may interact with one another in several ways: They may be in a producer/consumer, predator/prey, or parasite/host relationship. Or one organism may scavenge or decompose another. Relationships may be competitive or mutually beneficial. Some species have become so adapted to each other that neither could survive without the other.*

Science
Life science
- interdependence
- food webs
- biodiversity

Integrated Processes
Observing
Comparing and contrasting
Predicting
Inferring

Materials
40 pieces of red yarn, 4 feet each
2 pieces of green yarn, 4 feet each
Colored pencils
Labels on colored paper:

Source of Energy—Yellow
- *Sun*

Primary Producers—Green
- *Brazil Nut Tree #1000*
- *Bromeliad* (support)
- *Orchid* (support)

Primary Consumers—Blue
- *Cockroach* (debris)
- *Leaf-Cutter Ants* (leaf)
- *Termites* (wood)
- *Caterpillar* (leaf)
- *Capuchin Monkey* (leaf)
- *Macaw* (nuts) *#100*
- *Agouti* (nuts)
- *Machiguenga People* (nuts)
- *Flies*
- *Moths*
- *Butterflies*
- *Bees*
- *Millipedes*
- *Aquatic Insects*

Secondary Consumers—Orange
- *Tamandua*
- *Wasp*
- *Harpy Eagle #10*
- *Ocelot*
- *Margay*
- *Poison Dart Frog*

Tertiary Consumers—Red
- *Praying Mantis*
- *Beetle*

Background Information

Food webs exist in all habitats and can be used to demonstrate the complexity and energy flow in an ecosystem. Producers capture the sun's energy to make their own food in plant form, while consumers rely on eating those plants or other consumers to get their energy. At each feeding level, there is a 90% loss of energy that was available to the preceding level. Therefore, with each succeeding level having only 10% of the energy available, the number of individuals must decrease. This is why there are so few top level predators and so many low level (primary) consumers.

Comparing a rain forest food web to a deciduous forest food web shows just how biologically diverse or how full of different species a rain forest really is. For example, in a mere one-tenth of a hectare in the Choco region of Columbia, you can find 208 different tree species, or producers. In an entire forest of New Hampshire, you would find only 12 different tree species.* In addition to the number of trees, scientists believe that over 50% of all species live in the world's rain forests. That's an amazing fact considering rain forests occupy only about seven percent of the Earth's land mass. Terry Erwin, an entomologist with the Smithsonian Institution, discovered 1200 species of beetles on one tree in Panama. Of these 1200 species, he believes 160 live only on that kind of tree.**

A single Brazil Nut tree may be the primary producer for dozens of species, some specific to that single tree. Destroying the tree also destroys the species living on it. Many different consumers also eat different parts of the tree, partitioning resources to accommodate all needing food. Truly a complex food web on many levels, or NET WORTH SAVING!

There have been attempts to grow Brazil Nut trees in plantations; those attempts have failed. It was learned that the tree needs a Euglossine bee as a pollinator. This bee lives on a Euglossine orchid that grows on the surface of another tree. So, in order to successfully grow Brazil Nut trees, both trees must be planted. Note: In the ecology, there's a bee, another tree, an orchid, a macaw, and an agouti—all for the Brazil Nut tree's survival!

* Myers, Norman. *The Primary Source.* Norton. New York. 1992.

** Stephenson, Mary Ann and Peterson, Janet. *The Rainforest Connection II: A Curriculum Guide to Tropical Rainforests.* 5917 Hemstead Rd. Madison, WI 53711.

Management

1. You will need to have a large, open space such as a gymnasium, all-purpose room, etc., in order to make this food web. The activity can be done outdoors.
2. To make the labels for students to hold, print the names of the plants and animals (listed in italics under *Materials*) on the appropriately colored paper. The labels are color-coded. Yellow represents the source of energy—the sun; Green represents *Primary Producers;* Blue represents *Primary Consumers;* Orange represents *Secondary Consumers;* Red represents *Tertiary Consumers.* Be certain to include the numbers as indicated for the Brazil Nut tree, the macaw, and the harpy eagle. The numbers are representative of the energy at each level.
3. If your class has more students than there are labels, make extra *Insect* labels so that everyone can play a role. Most of the insects are likely to have three, four, or more other insects waiting in line to eat them (more partitioned resources). Thus, there may be four, five, or six more feeding levels. The *sloth* can be used if desired, if a connection to another tree is made.
4. Make certain that the students stay in their specific positions throughout the activity.
5. Students should be made to realize that the connections in this activity do not in any way represent all the possible connections in the food web. This is only a model of the diversity that exists.

Procedure

1. Distribute copies of the food web to each student. Have them use their colored pencils to color-code the different plants and animals on the sheet. (See *Materials* for the color coding.) These colors will correspond to the colored labels they will hold.
2. In order to realize how complex a rain forest is, explain to the students that they will be modeling a rain forest food web from *one single* Brazil Nut tree. Tell them that they will role-play the plants and animals in the web and will connect with yarn to the food source they eat.
3. Also, tell them they will be tracing the energy flow from the sun to the top level predators.
4. Explain that in addition to energy flow, they will be able to see that by eating different parts, more than one type of animal can live on one tree.
5. Direct the students to use their color-coded food web diagram as a guide to construct the class web.
6. Distribute the *Sun* label to a student and three pieces of yarn.
7. Distribute the *Brazil Nut Tree #1000,* the *Orchid,* and the *Bromeliad* labels to three students. Direct the three to each pick up one free end of yarn that the *Sun* is holding. (The sun provides energy for the Brazil Nut tree, the orchid, and the bromeliad enabling these plants to photosynthesize (make their own food). Explain that the yarn represents the flow of energy from the sun to the primary producers.
8. Give the *Brazil Nut Tree* 10 pieces of yarn, two of which will be green to represent support for the *Bromeliad* and the *Orchid* which are epiphytes

(a plant that grows off another plant but is not parasitic). These plants use the tree only as a perch. The other eight pieces of yarn will go to the animals (*Agouti, Machiguenga People, Macaw #100, Caterpillar, Capuchin Monkey, Termites, Leaf-Cutter Ants*) as seen on the diagram. Again remind the students that the yarn represents the energy flow.

9. Continue until the entire web is completed with students holding the labels and appropriate pieces of yarn.
10. For *Connecting Learning,* ask students to remain standing while holding the yarn pieces.

Connecting Learning

1. (Have the students hold up their hands with the pieces of yarn and look at the connections.) What do you see when you look at the connections? [It is not a simple food chain. This is very complex, more like a net.]
2. What plant or animal do you represent? What are you eating? What is eating you?
3. What we have represented is for one tree. What do you think it might look like if all the trees in a single acre of rain forest were represented?
4. How do all these organisms live on one Brazil Nut tree? [The macaw and Machiguenga people eat nuts. The agoutis pick up the nuts the macaws drop and cache them away, thereby spreading the tree's seeds. The monkey, caterpillars, and leaf-cutter ants all eat leaves in different sections of the tree. The cockroach eats leaves that have fallen and become debris, helping the decomposition process. With the help of the bacteria in their systems, the termites eat the wood. All are specialists that live on specialized diets.]
5. Why do you think chopping one Brazil Nut tree can be so devastating to so many species in the rain forest? [Destroying one tree is like destroying an island. The animals need to get to another tree or another island. This is hard to do if you cannot fly or hitch a ride on some larger animal that is able to move large distances.]
6. Why do you think someone would want to cut down a Brazil Nut tree? [lumber, firewood for cooking and warmth, to open up an area for farming, etc.]
7. Why do you think different colors are used on the labels? [They represent different levels in the food web.] Name those levels. (See *Management 2.*)
8. Explain the energy losses as you move from the producer to the consumers to the top level predator. Look at the drop in numbers. [Start with the Brazil Nut tree to represent the total number of producers at 1000. The total number of primary consumers would be about 100 (represented by the macaws). The total at the secondary consumer would drop 90% to 10 (represented by the harpy eagle). There are few eagles, so they are top level predators.]
9. The harpy eagle soars above many trees and eats consumers on those trees. Explain how the harpy eagle adds to the complexity of the rain forest food web. [Because it flies from tree to tree, it connects those trees together into an even more complex web.]
10. Because this pattern of netting represents only ONE way in which the interrelationships can be shown, describe other paths to connect the yarn.
11. What are you wondering now?

Curriculum Correlation

Cherry, Lynne. *The Great Kapok Tree: A Tale of the Amazon Rain Forest.* Harcourt Brace Jovanovich. New York. 1990.

Gibbons, Gail. *Nature's Green Umbrella.* HarperCollins Juvenile Books. Toronto. 1997.

Pratt, Kristin Joy. *A Walk in the Rainforest.* Dawn Publications. Nevada City, CA. 1992.

Silver, Donald M. *One Small Square: Tropical Rain Forest.* Learning Triangle Press. New York. 1999.

Stille, Darlene R. *Tropical Rain Forests.* Children's Press. Chicago. 2000.

What's the Net Worth?

Key Question

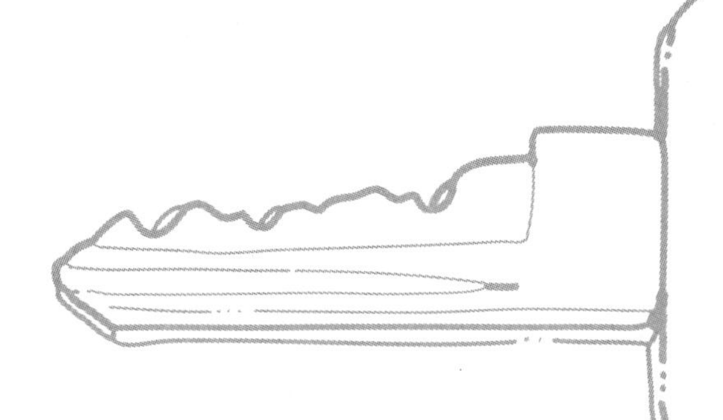

Using the Brazil Nut tree as an example, how would you explain the importance of protecting the trees in the rain forest?

Learning Goal

model the complex biodiversity, energy levels, and interdependence in a rain forest by role-playing the plants and animals in a rain forest food web.

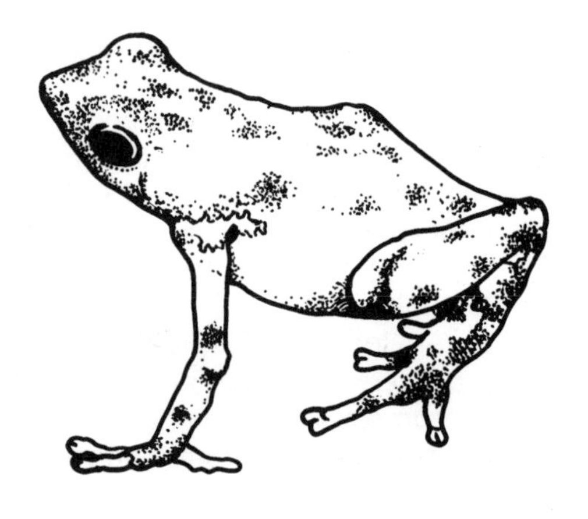

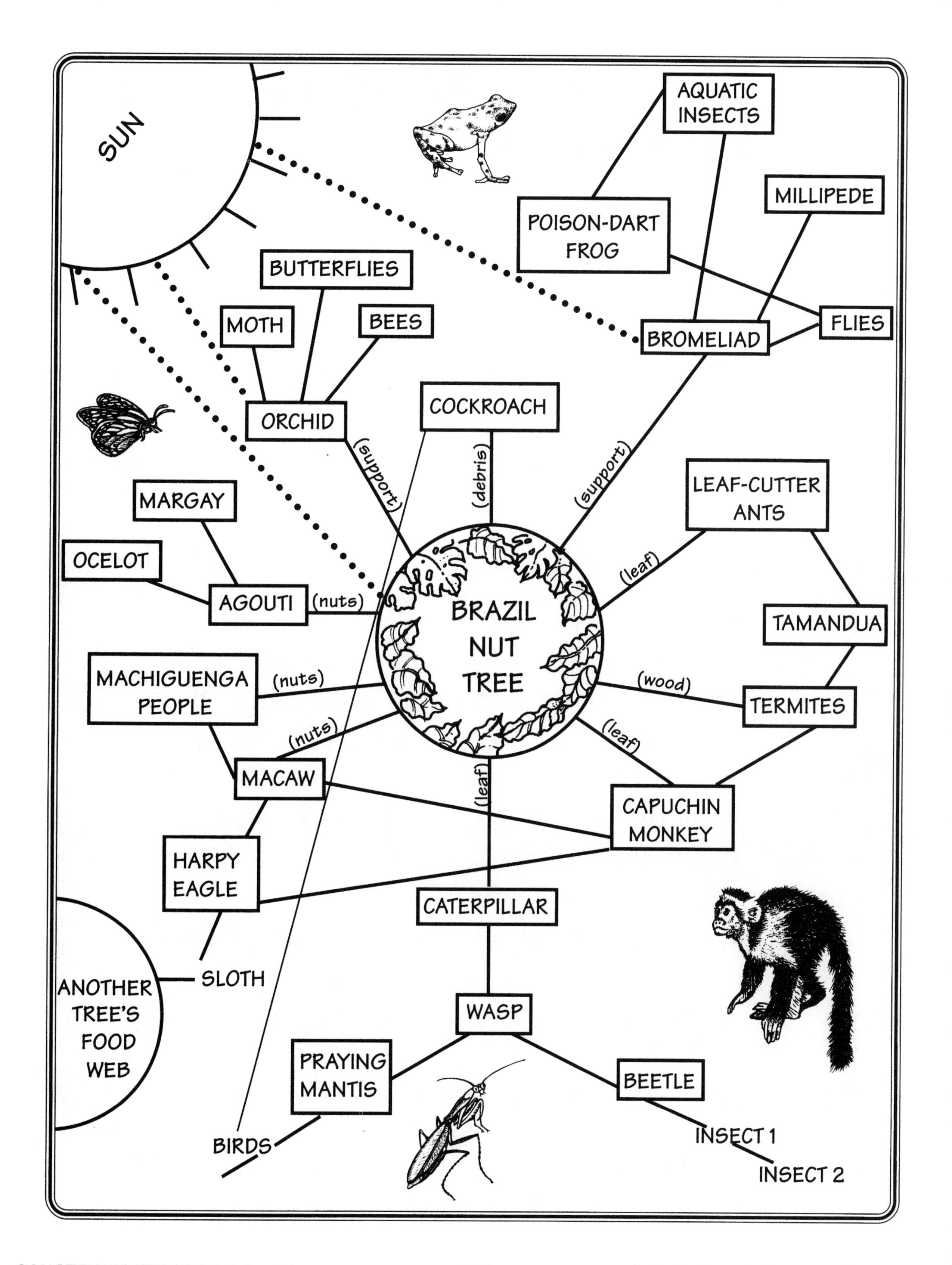
SUN
AQUATIC INSECTS
POISON-DART FROG
MILLIPEDE
BUTTERFLIES
MOTH
BEES
BROMELIAD
FLIES
ORCHID
COCKROACH
(support)
(debris)
(support)
LEAF-CUTTER ANTS
MARGAY
OCELOT
AGOUTI
(nuts)
(leaf)
BRAZIL NUT TREE
TAMANDUA
MACHIGUENGA PEOPLE
(nuts)
(wood)
TERMITES
(nuts)
(leaf)
MACAW
(leaf)
CAPUCHIN MONKEY
HARPY EAGLE
CATERPILLAR
SLOTH
ANOTHER TREE'S FOOD WEB
WASP
PRAYING MANTIS
BEETLE
BIRDS
INSECT 1
INSECT 2

What's the Net Worth?

CONNECTING LEARNING

Connecting Learning

1. What do you see when you look at the connections?

2. What plant or animal do you represent? What are you eating? What is eating you?

3. What we have represented is for one tree. What do you think it might look like if all the trees in a single acre of rain forest were represented?

4. How do all these organisms live on one Brazil Nut tree?

5. Why do you think chopping one Brazil Nut tree can be so devastating to so many species in the rain forest?

6. Why do you think someone would want to cut down a Brazil Nut tree?

What's the Net Worth?

CONNECTING LEARNING

Connecting Learning

7. Why do you think different colors are used on the labels?

8. Explain the energy losses as you move from the producer to the consumers to the top level predator. Look at the drop in numbers.

9. The harpy eagle soars above many trees and eats consumers on those trees. Explain how the harpy eagle adds to the complexity of the rain forest food web.

10. Because this pattern of netting represents only ONE way in which the interrelationships can be shown, describe other paths to connect the yarn.

11. What are you wondering now?

Concerning Critters: Adaptations & Interdependence

Materials List

Equipment

*Graduated cylinders, optional (#1915)
*Rulers (#1909)
*Thermometers (#1976)
*Hand lenses (#1977)
*Balances (#1917)
*Gram masses (#1923)
Meter stick
Forceps
Computer with projection system and Internet access

*Available from AIMS

Consumables and Non-consumables

Rubber bands—#19, various sizes
Black permanent marker
Fine-line markers
Colored pencils
Ink pads
Tempera paint
Transparent tape
Clear packing tape
Masking tape
Glue sticks
White glue
Stapler
Paper clips—jumbo and small
Scissors
Bulletin board paper
Chart paper
Construction paper, 12" x 18"—blue, red, black, white, brown, green
Copy paper—green, yellow, blue, orange, red
Cardboard
Transparency film
Index cards, 3" x 5"
Sticky notes, 3" and 1.5" x 2"
Newspaper want ads
Narrow-necked bottles
Empty two-liter bottles
Small bowls
Tubs of water
Shoeboxes
Egg cartons
Empty copy paper box
Plastic cups, 9 oz
Plastic cups, 16 oz
Paper plates, small (6") and large
Paper lunch sacks
Plastic sandwich bags
Paper towels
Plastic wrap
Zipper-type plastic bags, pint size
Flexible drinking straws
Regular drinking straws
Coffee stirrers
Styrofoam meat trays
Clear plastic gloves
Toothpicks
Thread
String
Yarn—brown, yellow, red, green
Moleskin
Cotton balls (at least 300)
1" wooden blocks
Clothespins
Sponge pieces
Velcro strips or dots
Party blowers
Small balloons
Chenille stems
Modeling clay
Miniature holiday lights, 50 or more
Owl pellets
Eggs
Large marshmallows
Miniature marshmallows
Multi-colored loop cereal, one box
Plain popped popcorn

Literature

The Wide-Mouthed Frog by Rex Schneider
Eyewitness Books: Amphibian by Barry Clarke
What if There Were No Bees? by Suzanne Slade
The Great Kapok Tree by Lynne Cherry

The AIMS Program

AIMS is the acronym for "**A**ctivities **I**ntegrating **M**athematics and **S**cience." Such integration enriches learning and makes it meaningful and holistic. AIMS began as a project of Fresno Pacific University to integrate the study of mathematics and science in grades K-9, but has since expanded to include language arts, social studies, and other disciplines.

AIMS is a continuing program of the non-profit AIMS Education Foundation. It had its inception in a National Science Foundation funded program whose purpose was to explore the effectiveness of integrating mathematics and science. The project directors, in cooperation with 80 elementary classroom teachers, devoted two years to a thorough field-testing of the results and implications of integration.

The approach met with such positive results that the decision was made to launch a program to create instructional materials incorporating this concept. Despite the fact that thoughtful educators have long recommended an integrative approach, very little appropriate material was available in 1981 when the project began. A series of writing projects ensued, and today the AIMS Education Foundation is committed to continuing the creation of new integrated activities on a permanent basis.

The AIMS program is funded through the sale of books, products, and professional-development workshops, and through proceeds from the Foundation's endowment. All net income from programs and products flows into a trust fund administered by the AIMS Education Foundation. Use of these funds is restricted to support of research, development, and publication of new materials. Writers donate all their rights to the Foundation to support its ongoing program. No royalties are paid to the writers.

The rationale for integration lies in the fact that science, mathematics, language arts, social studies, etc., are integrally interwoven in the real world, from which it follows that they should be similarly treated in the classroom where students are being prepared to live in that world. Teachers who use the AIMS program give enthusiastic endorsement to the effectiveness of this approach.

Science encompasses the art of questioning, investigating, hypothesizing, discovering, and communicating. Mathematics is a language that provides clarity, objectivity, and understanding. The language arts provide us with powerful tools of communication. Many of the major contemporary societal issues stem from advancements in science and must be studied in the context of the social sciences. Therefore, it is timely that all of us take seriously a more holistic method of educating our students. This goal motivates all who are associated with the AIMS Program. We invite you to join us in this effort.

Meaningful integration of knowledge is a major recommendation coming from the nation's professional science and mathematics associations. The American Association for the Advancement of Science in *Science for All Americans* strongly recommends the integration of mathematics, science, and technology. The National Council of Teachers of Mathematics places strong emphasis on applications of mathematics found in science investigations. AIMS is fully aligned with these recommendations.

Extensive field testing of AIMS investigations confirms these beneficial results:

1. Mathematics becomes more meaningful, hence more useful, when it is applied to situations that interest students.
2. The extent to which science is studied and understood is increased when mathematics and science are integrated.
3. There is improved quality of learning and retention, supporting the thesis that learning which is meaningful and relevant is more effective.
4. Motivation and involvement are increased dramatically as students investigate real-world situations and participate actively in the process.

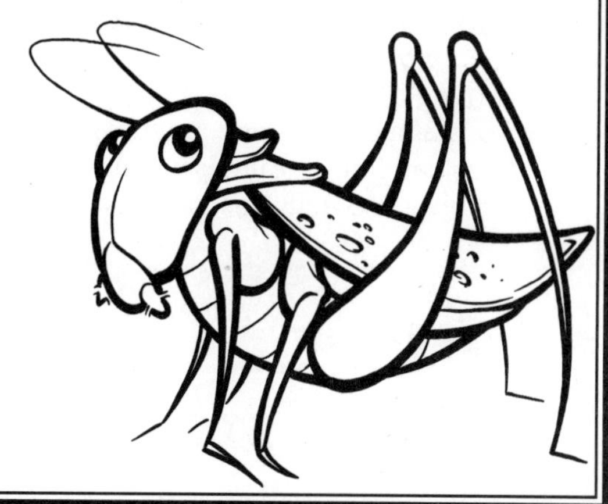

We invite you to become part of this classroom teacher movement by using an integrated approach to learning and sharing any suggestions you may have. The AIMS Program welcomes you!

Get the Most From Your Hands-on Teaching

When you host an AIMS workshop for elementary and middle school educators, you will know your teachers are receiving effective, usable training they can apply in their classrooms immediately.

AIMS Workshops are Designed for Teachers
- Hands-on activities
- Correlated to your state standards
- Address key topic areas, including math content, science content, and process skills
- Provide practice of activity-based teaching
- Address classroom management issues and higher-order thinking skills
- Include $50 of materials for each participant
- Offer optional college (graduate-level) credits

AIMS Workshops Fit District/Administrative Needs
- Flexible scheduling and grade-span options
- Customized workshops meet specific schedule, topic, state standards, and grade-span needs
- Sustained staff development can be scheduled throughout the school year
- Eligible for funding under the Title I and Title II sections of No Child Left Behind
- Affordable professional development—consecutive-day workshops offer considerable savings

Call us to explore an AIMS workshop
1.888.733.2467

Online and Correspondence Courses

AIMS offers online and correspondence courses on many of our books through a partnership with Fresno Pacific University.
- Study at your own pace and schedule
- Earn graduate-level college credits

See all that AIMS has to offer—visit us online

 **http://www.aimsedu.org**

Check out our website where you can:
- preview and purchase AIMS books and individual activities;
- learn about State-Specific Science and Essential Math;
- explore professional development workshops and online learning opportunities;
- buy manipulatives and other classroom resources; and
- download free resources including articles, puzzles, and sample AIMS activities.

find us on 

Become a fan of AIMS!
- Be the first to hear of new products and programs.
- Get links to videos on using specific AIMS lessons.
- Join the conversation—share how you and your students are using AIMS.

Hands-On Math and Science

AIMS FOR YOU

SIGN UP TODAY!

While visiting the AIMS website, sign up for our FREE *AIMS for You* e-mail newsletter to get free activities, puzzles, and subscriber-only specials delivered to your inbox monthly.

AIMS Program Publications

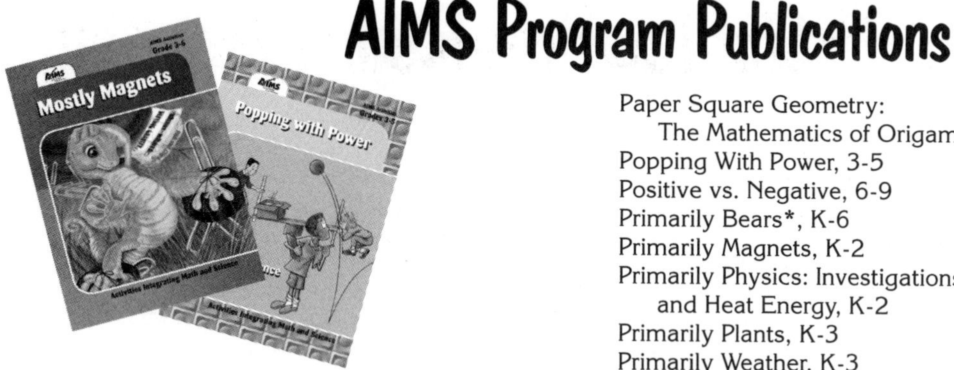

Actions With Fractions, 4-9
The Amazing Circle, 4-9
Awesome Addition and Super Subtraction, 2-3
Bats Incredible! 2-4
Brick Layers II, 4-9
The Budding Botanist, 3-6
Chemistry Matters, 5-7
Concerning Critters: Adaptations & Interdependence, 3-5
Counting on Coins, K-2
Cycles of Knowing and Growing, 1-3
Crazy About Cotton, 3-7
Critters, 2-5
Earth Book, 6-9
Earth Explorations, 2-3
Earth, Moon, Sun, 3-5
Earth Rocks! 4-5
Electrical Connections, 4-9
Energy Explorations: Sound, Light, and Heat, 3-5
Exploring Environments, K-6
Fabulous Fractions, 3-6
Fall Into Math and Science*, K-1
Field Detectives, 3-6
Floaters and Sinkers, 5-9
From Head to Toe, 5-9
Getting Into Geometry, K-1
Glide Into Winter With Math and Science*, K-1
Gravity Rules! 5-12
Hardhatting in a Geo-World, 3-5
Historical Connections in Mathematics, Vol. I, 5-9
Historical Connections in Mathematics, Vol. II, 5-9
Historical Connections in Mathematics, Vol. III, 5-9
It's About Time, K-2
It Must Be A Bird, Pre-K-2
Jaw Breakers and Heart Thumpers, 3-5
Looking at Geometry, 6-9
Looking at Lines, 6-9
Machine Shop, 5-9
Magnificent Microworld Adventures, 6-9
Marvelous Multiplication and Dazzling Division, 4-5
Math + Science, A Solution, 5-9
Mathematicians are People, Too
Mathematicians are People, Too, Vol. II
Mostly Magnets, 3-6
Movie Math Mania, 6-9
Multiplication the Algebra Way, 6-8
Out of This World, 4-8
Paper Square Geometry:
The Mathematics of Origami, 5-12
Popping With Power, 3-5
Positive vs. Negative, 6-9
Primarily Bears*, K-6
Primarily Magnets, K-2
Primarily Physics: Investigations in Sound, Light,
and Heat Energy, K-2
Primarily Plants, K-3
Primarily Weather, K-3
Probing Space, 3-5
Problem Solving: Just for the Fun of It! 4-9
Problem Solving: Just for the Fun of It! Book Two, 4-9
Proportional Reasoning, 6-9
Puzzle Play, 4-8
Ray's Reflections, 4-8
Sensational Springtime, K-2
Sense-able Science, K-1
Shapes, Solids, and More: Concepts in Geometry, 2-3
Simply Machines, 3-5
The Sky's the Limit, 5-9
Soap Films and Bubbles, 4-9
Solve It! K-1: Problem-Solving Strategies, K-1
Solve It! 2nd: Problem-Solving Strategies, 2
Solve It! 3rd: Problem-Solving Strategies, 3
Solve It! 4th: Problem-Solving Strategies, 4
Solve It! 5th: Problem-Solving Strategies, 5
Solving Equations: A Conceptual Approach, 6-9
Spatial Visualization, 4-9
Spills and Ripples, 5-12
Spring Into Math and Science*, K-1
Statistics and Probability, 6-9
Through the Eyes of the Explorers, 5-9
Under Construction, K-2
Water, Precious Water, 4-6
Weather Sense: Temperature, Air Pressure, and Wind, 4-5
Weather Sense: Moisture, 4-5
What on Earth? K-1
What's Next, Volume 1, 4-12
What's Next, Volume 2, 4-12
What's Next, Volume 3, 4-12
Winter Wonders, K-2

Essential Math
Area Formulas for Parallelograms, Triangles, and Trapezoids, 6-8
Circumference and Area of Circles, 5-7
Effects of Changing Lengths, 6-8
Measurement of Prisms, Pyramids, Cylinders, and Cones, 6-8
Measurement of Rectangular Solids, 5-7
Perimeter and Area of Rectangles, 4-6
The Pythagorean Relationship, 6-8
Solving Equations by Working Backwards, 7

* Spanish supplements are available for these books. They are only available as downloads from the AIMS website. The supplements contain only the student pages in Spanish; you will need the English version of the book for the teacher's text.

For further information, contact:
AIMS Education Foundation • P.O. Box 8120 • Fresno, California 93747-8120
www.aimsedu.org • 559.255.6396 (fax) • 888.733.2467 (toll free)

Duplication Rights

No part of any AIMS books, magazines, activities, or content—digital or otherwise—may be reproduced or transmitted in any form or by any means except as noted below.

Standard Duplication Rights

- A person or school purchasing AIMS activities (in books, magazines, or in digital form) is hereby granted permission to make up to 200 copies of any portion of those activities, provided these copies will be used for educational purposes and only at one school site.
- For a workshop or conference session, presenters may make one copy of any portion of a purchased activity for each participant, with a limit of five activities or up to one-third of a book, whichever is less.
- All copies must bear the AIMS Education Foundation copyright information.
- Modifications to AIMS pages (e.g., separating page elements for use on an interactive white board) are permitted only within the classroom or school for which they were purchased, or by presenters at conferences or workshops. Interactive white board files may not be uploaded to any third-party website or otherwise distributed. AIMS artwork and content may not be used on non-AIMS materials.

Standard duplication rights apply to activities received at workshops, free sample activities provided by AIMS, and activities received by conference participants.

Unlimited Duplication Rights

Unlimited duplication rights may be purchased in cases where AIMS users wish to:

- make more than 200 copies of a book/magazine/activity,
- use a book/magazine/activity at more than one school site, or
- make an activity available on the Internet (see below).

These rights permit unlimited duplication of purchased books, magazines, and/or activities (including revisions) for use at a given school site.

Activities received at workshops are eligible for upgrade from standard to unlimited duplication rights.

Free sample activities and activities received as a conference participant are not eligible for upgrade from standard to unlimited duplication rights.

State-Specific Science modules are licensed to one classroom/one teacher and are therefore not eligible for upgrade from standard to unlimited duplication rights.

Upgrade Fees

The fees for upgrading from standard to unlimited duplication rights are as follows.
For individual activities, the cost is $5 per activity per school site.
For Literature Links bundles, the cost is $12 per bundle per school site.
For books, the cost is based on the price of the book (see table).

Book Price	Upgrade Fee
$9.95	$15.00/site
$18.95	$24.00/site
$21.95	$27.00/site
$24.95	$30.00/site
$29.95	$35.00/site
$34.95	$40.00/site
$49.95	$55.00/site

The cost of upgrading is shown in the following examples:
For five activities at six schools:
5 activities x $5 x 6 schools = $150

For two books (at $21.95) at 10 schools:
2 books x $27 x 10 schools = $540

For three books (at $24.95) and four activities at eight schools:
(3 books x $30 x 8 schools) + (4 activites x $5 x 8 schools) = $720 + $160 = $880

Purchasing Unlimited Duplication Rights

To purchase unlimited duplication rights, please provide us the following:

1. The name of the individual responsible for coordinating the purchase of duplication rights.
2. The title of each book, activity, and/or magazine issue to be covered.
3. The number of school sites and name and address of each site for which rights are being purchased.
4. Payment (check, purchase order, credit card).

Requested duplication rights are automatically authorized with payment. The individual responsible for coordinating the purchase of duplication rights will be sent a certificate verifying the purchase.

Internet Use

AIMS materials may be made available on the Internet if all of the following stipulations are met:

1. The materials to be put online are purchased as PDF files from AIMS (i.e., no scanned copies).
2. Unlimited duplication rights are purchased for all materials to be put online for each school at which they will be used. (See above.)
3. The materials are made available via a secure, password-protected system that can only be accessed by employees at schools for which duplication rights have been purchased.

AIMS materials may not be made available on any publicly accessible Internet site.